Student Solutions Manual for

Physical Chemistry

Keith J. Laidler
University of Ottawa
Ottawa, Ontario

John H. Meiser
Ball State University
Muncie, Indiana

D1518289

The Benjamin/Cummings Publishing Company, Inc.
Menlo Park, California • Reading, Massachusetts
London • Amsterdam • Don Mills, Ontario • Sydney

ISBN 0-8053-5683-5

GHIJ-AL-8987

The Benjamin/Cummings Publishing Company, Inc.
2727 Sand Hill Road
Menlo Park, California 94025

TABLE OF CONTENTS

1 The Nature of Physical Chemistry and the Behavior of Gases....................................1

2 The First Law of Thermodynamics....................15

3 The Second and Third Laws of Thermodynamics.......33

4 Chemical Equilibrium..............................52

5 Phases and Solutions..............................73

6 Phase Equilibrium.................................88

7 Solutions of Electrolytes.........................107

8 Electrochemical Cells.............................123

9 Kinetics of Elementary Reactions.................138

10 Composite Reaction Mechanisms....................164

11 Quantum Mechanics and Atomic Structure...........185

12 The Chemical Bond................................207

13 Chemical Spectroscopy............................217

14 Molecular Statistics.............................238

15 The Solid State..................................261

16 The Liquid State.................................277

17 Surface Chemistry and Colloids...................285

18 Transport Properties.............................298

19 Macromolecules...................................311

CHAPTER 1: WORKED SOLUTIONS

THE NATURE OF PHYSICAL CHEMISTRY AND THE BEHAVIOR OF GASES

1. Work, $w = \int_{l_0}^{l} F(l)\ dl = \frac{1}{2} mu^2 - \frac{1}{2} mu_0^2$

   Speed = 88 km h^{-1} = 88(km h^{-1}) x (1/3600 s h^{-1}) x (10^3 m km^{-1}) = 24.4 m s^{-1}

   Rabbit $w = \frac{1}{2}(1000 \text{ kg})(24.4 \text{ m s}^{-1})^2 = 298$ kJ

   Monarch $w = \frac{1}{2}(1600 \text{ kg})(24.4 \text{ m s}^{-1})^2 = 476$ kJ

   The work required is directly proportional to the mass of the car.

2. If t^u is the value of the temperature,

   $28.1 = 27.5(1 + 0.160 \times 10^{-4}\ t^u + 0.10 \times 10^{-7}\ t^{u2})$

   which reduces to

   $(0.10 \times 10^{-7})t^{u2} + (0.160 \times 10^{-4})t^u - 0.022 = 0$

   The solution is t^u = 885 and therefore

   t = 885°C

3. $m_1 = 1 \times 10^{-24}$ g $\quad u_1 = 500$ m s^{-1}

   $m_2 = 1 \times 10^{-23}$ g $\quad u_2 = 0$

   Conservation of momentum: (1) $m_1u_1 + m_2u_2 = m_1u_1' + m_2u_2'$ (1)

   Conservation of energy: (2) $\frac{1}{2}m_1u_1^2 + \frac{1}{2}m_2u_2^2 = \frac{1}{2}m_1u_1'^2 + \frac{1}{2}m_2u_2'^2$ (2)

   $u_2 = 0$; from equation 1: $u_1' = u_1 - \frac{m_2}{m_1}\ u_2'$

   Substitute into Eq. 2 and solve for u_2':

$$u_2' = \frac{2u_1}{1 + \frac{m_2}{m_1}} = \frac{2(500 \text{ m) s}^{-1}}{1 + \frac{1 \times 10^{-23} \text{ g}}{1 \times 10^{-24} \text{ g}}} = 90.9 \text{ m s}^{-1}$$

$$\text{Kinetic energy} = \frac{1}{2} m_2 u_2'^2 = \frac{1}{2}(1 \times 10^{-26} \text{ kg}) \times (90.9)^2 \text{ m}^2 \text{ s}^{-2}$$

$$= 4.13 \times 10^{-23} \text{ J}$$

4. $$\text{Power} = \frac{dU}{dt} = \frac{8000 \times 10^3 \text{ J}}{24 \text{ h} \times 3600 \text{ s h}^{-1}}$$

$$= 92.6 \text{ J s}^{-1} = 92.6 \text{ watts}$$

5. A column of mercury 1 m^2 in cross-sectional area and exactly 0.760 m in height has a volume of 0.760 m^3 and a mass of 0.760 m^3 x 13 595.1 kg m^{-3}. Mass times the gravitational acceleration is a weight or force. The column's weight on the unit area gives a pressure

$$0.760 \text{ m}^3 \times 13\ 595.1 \text{ kg m}^{-3} \times 9.806\ 65 \text{ m s}^{-1}$$

$$= 101\ 325\ 0144 \text{ kg m s}^{-2}$$

Since 1 Pa = 1 kg m s^{-2}, the pressure is

$$= 101.325\ 0144 \text{ kPa}$$

6. $$P_1V_1 = P_2V_2$$

$$V_2 = \frac{1.80 \times 10^5 \text{ Pa} \times 0.30 \text{ dm}^3}{1.15 \times 10^5 \text{ Pa}}$$

$$= 0.47 \text{ dm}^3$$

7. $$\frac{V_1}{T_1} = \frac{V_2}{T_2}$$

$$V_2 = \frac{0.30 \text{ dm}^3}{330 \text{ K}} (550 \text{ K}) = 0.5 \text{ dm}^3$$

8. Concentration = $n/V = P/RT$. Since P

$= 1.013\ 25 \times 10^5$ Pa = 1 atm,

(a) $$\frac{n}{V} = \frac{1.013\ 25 \times 10^5 \text{ (Pa)}}{8.314 \text{ (J K}^{-1}\text{ mol}^{-1}\text{) } 298.15 \text{ (K)}}$$

$= 40.88 \text{ mol m}^{-3}$ since J/Pa = m^3

$= 0.0409 \text{ mol dm}^{-3}$

Number of molecules per unit volume

$= 0.0409 \text{ (mol dm}^{-3}\text{) } 6.022 \times 10^{23} \text{ mol}^{-1}$

$= 2.463 \times 10^{22} \text{ dm}^{-3}$

(b) $$\frac{n}{V} = \frac{1 \times 10^{-4} \text{ (Pa)}}{8.314\text{(J K}^{-1}\text{ mol}^{-1}\text{)}298.15 \text{ (K)}}$$

$= 4.03 \times 10^{-8} \text{ mol m}^{-3} = 4.03 \times 10^{-8} \text{ mol dm}^{-3}$

$= 2.429 \times 10^{16} \text{ dm}^{-3}$

9. (a) $\dfrac{V_1}{T_1} = C = \dfrac{V_2}{T_2}$ $\quad V_2 = \dfrac{2.50 \text{ dm}^3}{303 \text{ K}}(273.15 \text{ K}) = 2.25 \text{ dm}^3$

(b) Molar mass, $M = \dfrac{mRT}{PV}$

$$= \frac{1.08 \times 10^{-3}\text{(kg) } 8.314\text{(J mol}^{-1}\text{K}^{-1}\text{) } 273.15 \text{ (K)}}{(101.3 \times 10^3 \text{ Pa})(2.25 \times 10^{-4} \text{ m}^3)}$$

$= 0.1076 \text{ kg mol}^{-1} = 107.6 \text{ g mol}^{-1}$

10. $$M = \frac{\rho RT}{P} = \frac{1.92\text{(kg m}^{-3}\text{) } 8.314\text{(J mol}^{-1}\text{K}^{-1}\text{) } 298.15 \text{ K}}{150 \times 10^3 \text{ Pa}}$$

$= .0317 \text{ kg mol}^{-1} = 31.7 \text{ g mol}^{-1}$

11. $$n/V = P/RT = \frac{1.013\ 25 \times 10^5 \text{ (Pa)}}{8.314\text{(J K}^{-1}\text{ mol}^{-1}\text{)}298.15 \text{ (K)}}$$

$= 40.88 \text{ mol m}^{-3} = 0.040\ 80 \text{ mol dm}^{-3}$

Thus 0.040 88 mol = 1.159 g

1 mol = 28.4 g = molar mass

12. From Eq. 1.53

$$P_t = P_{H_2} + P_{H_2O}$$

$$P_{H_2} = P_t - P_{H_2O}$$

$$= 99.99 \text{ kPa} - 3.17 \text{ kPa} = 96.82 \text{ kPa}$$

13. The lifting force comes from the difference between the mass of air displaced and the mass of helium that replaces the air. Assume air to have an average molar mass of 28.8 g mol^{-1}. (80 mol % N_2 = 22.4 g mol^{-1} air; 20 mol % O_2 = 6.4 g mol^{-1} air)

Lifting force = $V(\rho_{air} - \rho_{helium})$ = 1000 kg

$$\rho_{air} = \frac{PM}{RT} = \frac{0.940(\text{atm})\, 101\,325(\text{Pa atm}^{-1})\, 28.8(\text{g mol}^{-1})}{8.314(\text{J K}^{-1}\text{ mol}^{-1})\, 290(\text{K})\ 10^3 \text{ g kg}^{-1}}$$

$$= 1.138 \text{ kg m}^{-3}$$

$$\rho_{He} = \frac{PM}{RT} = \frac{0.940(\text{atm})\, 101\,325(\text{Pa atm}^{-1})\, 4.003(\text{g mol}^{-1})}{8.314(\text{J K}^{-1}\text{ mol}^{-1})\ 290\ (\text{K})\ 10^3 \text{ g kg}^{-1}}$$

$$= 0.158 \text{ kg m}^{-3}$$

$$V = \frac{1000 \text{ kg}}{(1.138 - 0.158) \text{ kg m}^{-3}} = 1020 \text{ m}^3$$

14. The amount of butane present is given by $n = PV/RT$

$$n = \frac{101\,325 \text{ Pa} \times 40.0 \text{ dm}^3 \times 10^{-3} \text{ m}^3 \text{ dm}^{-3}}{8.314\ (\text{J K}^{-1} \text{ mol}^{-1})\ 298 \text{ K}}$$

$$= 1.64 \text{ mol}$$

The mixture required is 95 parts of argon to 5 parts of butane or 19 to 1.

n_{argon} = 19 x 1.64 = 31.1 mol

Mass of argon = 31.1 mol x 39.9 g mol^{-1} = 1240 g

The final pressure P_f is proportional to the total amount of gas; thus P_f = 1 atm x 20/1 = 20 atm

15. Rate of effusion is proportional to $1/T$; therefore from Graham's law

$$\frac{v_A}{v_B} = \frac{1}{2.3} = \sqrt{\frac{28\ \text{g mol}}{\text{molar mass of A}}}$$

Molar mass of A = $28\ (2.3)^2 = 148\ \text{g mol}^{-1}$

16. $PV = \frac{2}{3}\, nE_k;\ E_k = \frac{3}{2}\frac{PV}{n}$

$$E_k = \frac{3}{2}\left(\frac{1}{0.5\ \text{mol}}\right)(200\ \text{kPa})(8\ \text{dm}^3) = 4800\ \text{J mol}^{-1}$$

For one-half mole, E_k = 2400 J

17. (a) $n = \frac{PV}{RT} = \frac{152\ 000(\text{Pa})\,2(\text{dm}^3)\,10^{-3}(\text{m}^3\text{dm}^{-3})}{8.314\ (\text{J K}^{-1}\ \text{mol}^{-1})\,298.15\ (\text{K})}$

$= 0.1226$ mol

(b) The number of molecules = amount of substance x Avogadro constant

$= 0.1226\ (\text{mol}) \times 6.022 \times 10^{23}(\text{mol}^{-1})$

$= 7.385 \times 10^{22}$

(c) Rearrangement of Eq. 1.43 gives

$$\overline{u}^2 = \frac{3RT}{mL} \quad \text{or} \quad \sqrt{\overline{u}^2} = \sqrt{\frac{3RT}{M}}$$

$$\sqrt{\overline{u}^2} = \sqrt{\frac{3(8.314\ \text{J K}^{-1}\ \text{mol}^{-1})(298.15\ \text{K})}{0.028\ 134\ \text{kg mol}^{-1}}}$$

$= 515.2\ \text{m s}^{-1}$

(d) $\overline{\varepsilon}_k = \frac{1}{2}\, m\, \overline{u}^2 = \frac{0.028\ 0134\ (\text{kg mol}^{-1})(515.2)^2(\text{m}^2\ \text{s}^{-1})}{2(6.022 \times 10^{23}\ \text{mol}^{-1})}$

$= 6.174 \times 10^{-21}$ J

(e) E_k (total) = 0.1226 (mol) 6.022 x 10^{23} (mol^{-1}) x

6.174 x 10^{21} (J $molecule^{-1}$)

= 455.8 J

Alternatively,

E_k (total) = 0.1226 x $\frac{3}{2}$ RT

= (0.1226 mol)(3)(8.314 J K^{-1} mol^{-1})(298.15 K)/2

= 455.8 J

18. From Eq. 1.43 we have

$$\frac{\sqrt{\overline{u}_2^2}}{\sqrt{\overline{u}_1^2}} = \sqrt{\frac{T_2}{T_1}} = \sqrt{\frac{400}{300}} = \sqrt{1.333} = 1.15$$

19. For use in the mean-free-path equation, Eq. 1.72, the value of N_A, the number of molecules formed, must be calculated from the ideal gas law. The number of molecules per unit volume is

$$\frac{N_A}{V} = \frac{LP}{RT}$$

$$= \frac{6.022 \times 10^{23}(\text{mol}^{-1})\,101\ 325(\text{Pa})}{8.314\,(\text{J K}^{-1}\ \text{mol}^{-1})\ 298.15\,(\text{K})} = 2.46 \times 10^{25}\ \text{m}^{-3}$$

Since 1 Pa ≡ kg m^{-1} s^{-2} and J = kg m^2 s^{-2}

$\lambda = 1/(\sqrt{2}\pi d^2 N_A) = 1/(\sqrt{2}(\pi)(3.74 \times 10^{-10}\ \text{m})^2 \times$

$(2.46 \times 10^{25}\ \text{m}^{-3})$

$= 6.54 \times 10^{-8}$ m

$Z_A = \sqrt{2}\ \pi d^2\ \overline{u}_A N_A$

$= \sqrt{2}\ \pi(3.74 \times 10^{-10}\ \text{m})^2\ 474.6(\text{m s}^{-1})2.46 \times 10^{25}\text{m}^{-3}$

$= 7.26 \times 10^9\ \text{s}^{-1}$

$$Z_{AA} = \frac{1}{\sqrt{2}}\ \pi d^2 \bar{u}_A N_A^2 \ = \frac{1}{2}\ N_A Z_A$$

$$= 8.9 \times 10^{34}\ m^{-3}\ s^{-1}$$

20. $\lambda = \dfrac{V}{\sqrt{2}\ \pi d^2 N}$ where N = number of molecules

Since $PV = nRT = \dfrac{N}{L}RT$

$$\lambda = \frac{RT}{\sqrt{2}\ \pi d^2 LP}$$

21. (a) At 133.32 Pa,

$$\lambda = \frac{(8.314\ J\ K^{-1}\ mol^{-1})(298.15\ K)}{\sqrt{2}\pi(0.258 \times 10^{-9} m)^2(6.022 \times 10^{23} mol^{-1})} \times$$

$$1/(133.32\ Pa)$$

$$= 1.044 \times 10^{-4}\ m \text{ since } J/Pa = m^3$$

(b) At 101.325 kPa, $\lambda = 1.37 \times 10^{-7}$ m

Alternatively λ is lowered by a ratio of

$$\frac{101\ 325}{133.32} = 760$$

Thus $1.044 \times 10^{-4}/760 = 1.37 \times 10^{-7}$ m

(c) At 1.0×10^8 Pa,

$$\lambda = 1.044 \times 10^{-4}\ m \times \frac{133.32}{1.0 \times 10^8} = 1.39 \times 10^{-10}\ m$$

22. From Eq. 1.72

$$\lambda = \frac{V}{\sqrt{2}\ \pi d^2 N} = \frac{1\ m^3}{\sqrt{2}\ \pi(2.5 \times 10^{-10})^2 m^2}$$

$$= 3.60 \times 10^{18}\ m$$

This is about a hundred times greater than the distance between the earth and the nearest star (Proxima Centauri)!

23. The curves are similar to those in Figure 1.10

24. Ideal gas prediction: $P = nRT/V$

$$P = \frac{17.5 \text{ g}/2(35.45 \text{ g mol}^{-1})\, 8.314(\text{J mol}^{-1}\text{K}^{-1})\, 273.15 \text{ K}}{0.8 \times 10^{-3} \text{ m}^3}$$

$$= 700.67 \text{ kPa} = 6.91 \text{ atm}$$

Van der Waals prediction: $P = \dfrac{nRT}{V-nb} - \dfrac{an^2}{V^2}$

$n = 0.2468;\ a = 6.493;\ b = 0.0562$

$$P = \frac{0.2468 \text{ (g mol}^{-1})\, 0.082(\text{atm dm}^3 \text{ K}^{-1}\text{mol}^{-1})\, 273.15 \text{ K}}{0.8 \text{ (dm}^3) - (0.2468 \text{ mol})(0.0562 \text{ dm}^3 \text{ mol}^{-1})} - \frac{6.493(0.2468)^2}{(0.8)^2} \text{ atm}$$

$$= (7.032 - 6.18 \times 10^{-1}) \text{ atm} = 6.41 \text{ atm}$$

25. $T_r = \dfrac{356}{36.5 + 273.15} = 1.15;\ P_r = \dfrac{54}{71.7} = 0.75$

From Figure 1.13, $Z = \dfrac{PV}{RT} = 0.815$

Using the data of the problem

$$V = \frac{ZRT}{P} = \frac{0.815 \text{ (0.082 atm dm}^3 \text{ K}^{-1} \text{ mol}^{-1}) \text{ 356 K}}{54 \text{ atm}}$$

$$= 0.44 \text{ dm}^3 \text{ mol}^{-1}$$

26. The Dieterici equation is $P = \dfrac{RT}{V_m - b} e^{-a/RTV_m}$

$$\left(\frac{\partial P}{\partial V_m}\right)_T = RTe^{-a/RTV_m}\left[\frac{a}{V_m^2 RT(V_m-b)} - \frac{1}{(V_m-b)^2}\right]$$

$$= P\left[\frac{a}{V_m^2 RT} - \frac{1}{V_m-b}\right]$$

$$\left(\frac{\partial^2 P}{\partial V_m^2}\right)_T = P\left[-\frac{2a}{V_m^3 RT} + \frac{1}{(V-b)^2} + \left(\frac{a}{V^2 RT} - \frac{1}{(V-b)}\right)^2\right]$$

At the critical point $(\partial P/\partial V)_T = 0$, and by substituting the critical constants for P, V, and T, we have

(1) $$P_c\left[\frac{a}{V_c^2RT_c} - \frac{1}{(V_c-b)}\right] = 0$$

and similarly for the second dervitive which corresponds to the highest temperature a horizontal line may exist on the curve:

(2) $$P_c\left[-\frac{2a}{V_c^3RT_c} + \frac{1}{(V_c-b)^2} + \left(\frac{a}{V_c^2RT_c} - \frac{1}{(V_c-b)}\right)^2\right] = 0$$

After eliminating P_c from both expressions, a is obtained from equation (1): $a = V_c^2RT_c/(V_c-b)$

Substitution into equation 2 yields

$$\frac{-2V_c^2RT_c}{(V_c-b)(V_c^3RT_c)} + \frac{1}{(V_c-b)^2} + \left(\frac{V_c^2RT_c}{(V_c-b)V_c^2RT_c} - \frac{1}{(V_c-b)}\right)^2$$

$= 0$

$$\frac{-2}{V_c} + \frac{1}{V_c-b} = 0 \qquad b = V_c/2$$

Substitution of b into equation 1 yields

$a = 2V_cRT_c$

Therefore at the critical point

$$P_c = \frac{2RT_c}{V_c}e^{-2}$$

27. Write the van der Waals equation (Eq. 1.75) in the form

$$Z = \frac{PV}{RT} = 1 + \frac{Pb}{RT} - \frac{a}{RTV} + \frac{ab}{RTV^2}$$

When the pressure is low, $\lim_{P\to 0}\frac{ab}{RTV^2} = \lim_{P\to\infty}\frac{ab}{RTV^2} = 0$

In the term $\frac{a}{RTV}$, V may be replaced by $\frac{RT}{P}$, and we have

$$Z = \frac{PV}{RT} = 1 + \frac{Pb}{RT} - \frac{a}{RT}\left(\frac{P}{RT}\right) = 1 + \frac{P}{RT}\left(b - \frac{a}{RT}\right)$$

As P goes to zero, $Z = \frac{PV}{RT} = 1$

28. For a: 1 atm = 101 325 Pa, $(1\ \text{dm})^6 = (0.1\ \text{m})^6$

$$= 1 \times 10^{-6}\ \text{m}$$

$$a = 5.49\ \text{atm dm}^6\ \text{mol}^{-2} \times \frac{101\ 325\ \text{Pa}}{\text{atm}} \times \frac{1 \times 10^{-6}\ \text{m}}{\text{dm}^3}$$

$$= 0.5563\ \text{Pa m}^6\ \text{mol}^{-2}$$

For b: $(1\ \text{dm})^3 = (0.1\ \text{m})^3 = 10^{-3}\ \text{m}^3$

$$b = 0.0638\ \text{dm}^3\ \text{mol}^{-1} \times \frac{10^{-3}\text{m}^3}{\text{dm}^3}$$

$$= 0.0638 \times 10^{-3}\ \text{m}^3\ \text{mol}^{-1}$$

29. Ideal gas equation:

$$P = \frac{nRT}{V} = \frac{3(\text{mol})\,8.314(\text{J K}^{-1}\ \text{mol}^{-1})\,298.15\ (\text{K})}{0.008\ 25\ \text{m}^3}$$

$$= 902\ 400\ \text{Pa} = 8.91\ \text{atm}$$

Van der Waals equation:

$$P = \frac{RT}{V_m - b} - \frac{a}{V_m^2}$$

$$= \frac{8.314(\text{J K}^{-1}\ \text{mol}^{-1})\,298.15\ (\text{K})}{2.75 \times 10^{-3}\ \text{m}^3\ \text{mol}^{-1} - 4.27 \times 10^{-5}\ \text{m}^3\ \text{mol}^{-1}} - \frac{0.3637\ \text{Pa m}^6\ \text{mol}^{-2}}{(2.75 \times 10^{-3}\ \text{m}^3\ \text{mol}^{-1})^2}$$

$$= 867\ 500\ \text{Pa} = 8.56\ \text{atm}$$

Dieterici equation:

$$P = \frac{RT\ \exp(-a/V_mRT)}{V_m - b}$$

$$P = \frac{8.314(\text{J K}^{-1}\ \text{mol}^{-1})\,298.15\ (\text{K})}{2.75 \times 10^{-3}\ (\text{m}^3\ \text{mol}^{-1}) - 4.63 \times 10^{-5}\ (\text{m}^3\ \text{mol}^{-1})} \times$$

$$\exp(-0.462/2.75 \times 10^{-3} \times 8.314 \times 298.15)$$

$$= 856\ 200\ \text{Pa} = 8.45\ \text{atm}$$

Beattie-Bridgeman equation:

$$P = \frac{RT}{V_m^2}\left(1 - \frac{c}{V_m T^3}\right)\left(V_m + B_o - \frac{bB_o}{V_m}\right) - \frac{A_o}{V_m^2}\left(1 - \frac{a}{V_m}\right)$$

$$P = \frac{8.314(298.15)}{(0.002\ 75)^2}\left(1 - \frac{660}{(298.15)^3(0.002\ 75)}\right) \times$$

$$\left(0.002\ 75 + 104.76 \times 10^{-6} - \frac{72.35 \times 10^{-6}(104.76 \times 10^{-6})}{0.002\ 75}\right) - \frac{0.507\ 28}{(0.002\ 75)^2} \times$$

$$\left(1 - \frac{71.32 \times 10^{-6}}{0.002\ 75}\right)$$

$$861\ 027\ \text{Pa} = 8.50\ \text{atm}$$

All the calculated values for P are approximately the same under these conditions.

30. From Eq. 1.91, $b = V_c/3$ and $V_c = \frac{3RT_c}{8P_c}$, then

$$b = \frac{RT_c}{8P_c}$$

$$b = \frac{8.314(\text{J K}^{-1}\ \text{mol}^{-1})\ 473\ \text{K}}{8(30\ \text{atm} \times 101\ 325\ \text{Pa atm}^{-1})}$$

$$= 0.162 \times 10^{-3}\ \text{m}^3\ \text{mol}^{-1}$$

31. The Dieterici equation can be expanded in terms of the infinite series $e^x = 1 + x + x^2/2! + x^3/3! \cdots$.

This gives

$$P = \frac{RT}{(V_m - b)} - \frac{a}{V_m(V_m - b)} + \frac{a^2}{2\,RT\,V_m^2(V_m - b)} - \cdots$$

which is in the virial form. The second coefficient is $-\frac{a}{V_m - b}$. At low densities the third and higher terms are negligible and $V_m(V_m - b) \approx V_m^2$. Dropping the third term and substituting we obtain

$$P = \frac{RT}{V_m - b} - \frac{a}{V_m^2}$$

This is identically the van der Waals equation.

32. From Eq. 1.15 we have

$$\theta_{Hg} = \frac{(V_{50} - V_0)_{Hg}}{(V_{100} - V_0)_{Hg}} \times 100$$

Integrating the expression for α we have

$$V_{50} - V_0 = \int_0^{50} \alpha V_0 \, d\theta$$

$$= \left[1.817 \times 10^{-4}\ \theta\Big|_0^{50} + 2.95 \times 10^{-8}\ \theta^2\Big|_0^{50} + 1.15 \times 10^{-10}\ \theta^3\Big|_0^{50}\right] V_0$$

$$= 0.009\ 107\ V_0$$

$$V_{100} - V_0 = \left[1.817 \times 10^{-4}\theta\Big|_0^{100} + 2.95 \times 10^{-9}\ \theta^2\Big|_0^{100} + 1.15 \times 10^{-10}\ \theta^3\Big|_0^{100}\right] V_0 = 0.018\ 31\ V_0$$

Then

$$\theta_{Hg} = \frac{0.009\ 107\ V_0}{0.018\ 31\ V_0} \times 100 = 49.7°\text{C}$$

33. (a) The standard gravitational acceleration is $9.807\ \text{m s}^{-2}$. If g decreases by 0.010 m per km of height, this is equivalent to a change of

$0.010 \text{ m s}^{-2}/10^3 \text{ m} = 10^{-5} \text{ s}^{-2} z$ where z is the altitude. Therefore, $g = 9.807 \text{ m s}^{-2} - 10^{-5} \text{ s}^{-2} z$ and substitution in Eq. 1.82 gives

$$\frac{dP}{P} = -\frac{M}{RT}(9.807 - 10^{-5} z)\,dz$$

or

$$\ln P/P_o = -\frac{M}{RT}(9.807 \text{ m s}^{-2} z - 5 \times 10^{-6} \text{ s}^{-2} z^2)$$

(b) $$\ln P/P_o = -\frac{28.0(\text{g mol}^{-1})(10^{-3} \text{ kg g}^{-1})}{8.314 (\text{J K}^{-1} \text{ mol}^{-1}) 298.1 \text{ K}} \times$$

$$[9.807(10^5)\text{m}^2 \text{ s}^{-2} - 5 \times 10^{-6}(10^5)^2 \text{ m}^2 \text{ s}^{-2}]$$

$$\ln P/P_o = -10.51; \; P/P_o = 2.73 \times 10^{-5}$$

$$P = 2.73 \times 10^{-5} \text{ atm}$$

34. Consider the gamboge particles to be gas molecules in a column of air. Since the number of particles present is proportional to their pressure we write Eq. 1.79 as $dN/N = -(Mg/RT)dz$ or

$$\ln\frac{N}{N_o} = -\frac{Mg\Delta z}{RT}$$

$$RT \ln\frac{N}{N_o} = -mLg\Delta z \qquad \text{since } M = mL$$

The mass is $\frac{4}{3}\pi r^3 \Delta\rho$, and substitution gives

$$RT \ln N_o/N = \frac{4}{3}\pi r^3 Lg\Delta\rho\Delta z$$

$$8.314(\text{J K}^{-1} \text{ mol}^{-1})288.15(\text{K}^{-1})\ln\frac{100}{47}$$

$$= L\frac{4}{3}(3.1416)(2.12 \times 10^{-7} \text{ m})^3 \, 9.81 \text{ m s}^{-2} \times$$

$$(35 - 5) \times 10^{-6} \text{ m } (1.206 - 0.999)\text{g cm}^{-3} \times$$

$$\frac{\text{kg } 10^6 \text{ cm}^3}{10^3 \text{ g m}^3}; \text{then, } L = 7.44 \times 10^{23} \text{ mol}^{-1}$$

35. (a) A column of mercury 1 m^2 in cross-sectional area and 0.001 m in height has a volume of 0.001 m^3 and a mass of 0.001 m^3 x 13 595.1 kg m^{-3}. Then 1 mmHg = 0.001 m^3 x 13 595.1 kg m^{-1} x 9.806 65 m s^{-2}

= 133.322 3874 Pa

By definition 1 atmosphere = 101 325 Pa. Then

$$1 \text{ torr} = \frac{1}{760}(101\ 325 \text{ Pa}) = 133.322\ 3684 \text{ Pa}$$

Thus

$$1 \text{ mmHg} = \frac{133.322\ 3874}{133.322\ 3684} = 1.000\ 000\ 14 \text{ torr}$$

(b) From $PV = nRT$, and since $m = \frac{N}{L}$

$PV = \frac{NRT}{L}$ and the number density is $\frac{N}{V} = \frac{PL}{RT}$

For $P = 10^{-6}$ torr:

$$N/V = \frac{10^{-6} \text{ torr (1 atm/760 torr)}}{8.314 \text{ J K}^{-1} \text{ mol}^{-1}\ 298.15 \text{ K}} \times$$

$$101\ 325(\text{Pa atm}^{-1})6.022 \times 10^{23} \text{ mol}^{-1}$$

$$= 3.24 \times 10^{16} \text{ m}^{-3}$$

For $P = 10^{-15}$ torr:

$$N/V = \frac{10^{-15} \text{ torr(1 atm/760 torr)}}{8.314 \text{ (J K}^{-1} \text{ mol}^{-1})\ 298.15 \text{ K}} \times$$

$$101\ 325(\text{Pa atm}^{-1})6.022 \times 10^{23} \text{ mol}^{-1}$$

$$= 3.24 \times 10^{7} \text{ m}^{-3}$$

This is still a substantial number.

CHAPTER 2: WORKED SOLUTIONS
THE FIRST LAW OF THERMODYNAMICS

1. Weight of bird = 1.5 x 9.81 kg m s^{-2}

   Work required to raise it 75 m = 1.5 x 9.81 x 75 kg m^2 s^{-1}

   = 1103.6 J

   Kinetic energy = $\frac{1}{2}$ mv^2

   = $\frac{1}{2}$ x 1.5 x 20^2 = 300 J

   Total energy = 1103.6 + 300 = 1403.6 J

   = 1.404 kJ

2. If there are n mol initially,

   $$P_1V_1 = nRT$$

   Finally, P_2V_2 = (n + 0.27 mol) RT

   $\Delta(PV)$ = 0.27 RT

   ΔH = ΔU + $\Delta(PV)$

   = 9400 + (0.27 x 8.314 x 300) J

   = 9400 + 673 = 10 073 J

   = 10.07 kJ

3. 1 mol of ice has a volume of 18.01 g/0.9168 g cm^{-3}

   = 19.64 cm^3

   1 mol of water has a volume of 18.01 g/0.9998 g cm^{-3}

   = 18.01 cm^3

   ΔV(ice → water) = -1.63 cm^3 mol^{-1}

   $\Delta(PV)$ = -0.001 63 atm dm^3 = -0.165 J mol^{-1}

   ΔH = 6025 J = ΔU + $\Delta(PV)$ = ΔU - (0.165 J mol^{-1})

   ΔU = 6025 J mol^{-1} = 6.025 kJ mol^{-1}

(difference between ΔH and ΔU is only $0.165\ J\ mol^{-1}$)

Work done on the system = $0.165\ J\ mol^{-1}$

4. 1 mol of water at 100°C has a volume of $\dfrac{18.01\ g}{0.9584\ g\ cm^{-3}}$

$= 18.79\ cm^3$

Vol. of steam = $\dfrac{18.01\ g}{0.000\ 596\ g\ cm^{-3}} = 30\ 218.1\ cm^3$

Volume increase = $30\ 199\ cm^3\ mol^{-1}$

$= 30.20\ dm^3\ mol^{-1}$

$$\Delta(PV) = 30.20\ atm\ dm^3\ mol^{-1} = 3059.3\ J\ mol^{-1}$$

$$= 3.06\ kJ\ mol^{-1}$$

$$\Delta H = 40\ 627\ J\ mol^{-1} = \Delta U + (3059.3\ J\ mol^{-1})$$

$$\Delta U = 37.57\ kJ\ mol^{-1}$$

Work done by the system = $-w$ = 3.06 kJ

5. (1) Heat the water from -10°C to 0°C:

$$q_1 = C_P\ dT = C_P(T_2 - T_1) = 753\ J\ mol^{-1}$$

(2) Freeze the water at 0°C:

$$q_2 = 6025\ J\ mol^{-1}$$

(3) Cool the ice from 0°C to -10°C:

$$q_3 = -377\ J\ mol^{-1}$$

Net $q = 753 - 6025 - 377 = -5649\ J\ mol^{-1}$

$= \Delta H = -5.65\ kJ\ mol^{-1}$

6. Heat evolved = 1.69 x 6937 = 11 724 J

Molar mass of CH_3COCH_3 = 3 x 12.01 + 6 x 1.008 + 16.00

$= 58.08\ g\ mol^{-1}$

Heat evolved in the combustion of 1 mol

$$= \frac{11\ 724 \times 58.08}{0.70}\ \text{J} = 972.7\ \text{kJ}$$

(a) $\Delta U = -972.7\ \text{kJ mol}^{-1}$

(b) $CH_3COCH_3(\ell) + 4O_2(g) \rightarrow 3CO_2(g) + 3H_2O(\ell)$

$\Delta n(\text{gases}) = 1$

$\Delta H = -972\ 700 - 8.314 \times 299.8\ \text{J mol}^{-1}$

$= -975.2\ \text{kJ mol}^{-1}$

7. Molar mass of benzene $= 6 \times 12.01 + 6 \times 1.008$

$= 78.11\ \text{g mol}^{-1}$

Heat evolved in the combustion of 1 mol

$$= \frac{26.54 \times 78.11}{0.633} = 3274.9\ \text{kJ}$$

(a) $\Delta U = -3274.9\ \text{kJ mol}^{-1}$

(b) $C_6H_6(\ell) + \frac{15}{2}O_2(g) \rightarrow 6CO_2(g) + 3H_2O(\ell)$

$\Delta\nu(\text{gases}) = 1.5$

$\Delta H = -3\ 274\ 900 - 1.5 \times 8.314 \times 298.15\ \text{J mol}^{-1}$

$= -3278.6\ \text{kJ mol}^{-1}$

8. (a) Heat capacity of man $= 292\ 600\ \text{J K}^{-1}$

Temperature rise $= \frac{10\ 460\ 000\ \text{J}}{292\ 600\ \text{J K}^{-1}} = 35.7°\text{C}$

Final temperature $= 37 + 35.7 = 72.7°\text{C}$

(b) $\Delta H = 43\ 400\ \text{J mol}^{-1}/18\ \text{g mol}^{-1} = 2411\ \text{J g}^{-1}$

Mass of water required $= \frac{10\ 460\ 000\ \text{J}}{2411\ \text{J g}^{-1}} = 4340\ \text{g}$

$= 4.34\ \text{kg}$

9. The reaction is

$$Zn + H_2SO_4 \rightarrow ZnSO_4 + H_2(g)$$

Thus 1 mol of gas is liberated by each mole of Zn, i.e. by 65.37 g. One hundred grams therefore liberates (110/65.37) mol = 1.53 mol of H_2. The work done by the system is $P\Delta V$:

$$-w = P\Delta V = n_{H_2}RT$$
$$= 1.53 \text{ mol} \times 8.314 \text{ J K}^{-1} \text{ mol}^{-1} \times 298.15 \text{ K}$$
$$= 3790 \text{ J} = 3.79 \text{ kJ}$$

The work in a sealed vessel ($\Delta V = 0$) is zero.

10. The volume increase is

$$2.5 \times 10^{-6} \text{ m}^2 \times 2.4 \times 10^{-2} \text{ m} = 6.0 \times 10^{-8} \text{ m}^3$$

The work done by the system is $P\Delta V$:

$$-w = P\Delta V = 1.013 \times 10^5 \text{ Pa} \times 6.0 \times 10^{-8} \text{ m}^3$$
$$= 6.08 \times 10^{-3} \text{ J} \quad [\text{Pa} \equiv \text{N m}^{-2}; \text{ N m} \equiv \text{J}]$$

11. One gram of water is 1 g/18.02 g mol^{-1} = 0.0555 mol. One calorie is 4.184 J. The heat capacity is thus

$$\frac{1 \text{ cal}}{1 \text{ g} \times 1 \text{ K}} = \frac{4.184 \text{ J}}{0.0555 \text{ mol} \times 1 \text{ K}} = 75.4 \text{ J K}^{-1} \text{ mol}^{-1}$$

12. One kilogram of water is 55.5 mol. The temperature range is 75°K and the heat required is therefore

$$= 55.5 \text{ mol} \times 75 \text{ K} \times 75.4 \text{ J K}^{-1} \text{ mol}^{-1}$$
$$= 313\,900 \text{ J} = 314 \text{ kJ.}$$

A 1 kW heater supplies this energy in 314 s [J = W s].

13. $\Delta H° = 84.68 + 2\ (74.81) = 64.9\ \text{kJ mol}^{-1}$

14. Molar mass of $CH_3OH = 12.01 + 4 \times 1.008 + 16.00$

$$= 32.04\ \text{g mol}^{-1}$$

Amount of methanol = $5.27\ \text{g}/32.04\ \text{g mol}^{-1}$

Heat evolved = $\dfrac{119.50\ \text{kJ} \times 32.04\ \text{g mol}^{-1}}{5.27\ \text{g}}$

$$= 726.5\ \text{kJ mol}^{-1} = -\Delta_c U°$$

(a) $\Delta_c U° = -726.5\ \text{kJ mol}^{-1}$

$$CH_3OH\ (\ell) + \tfrac{3}{2}O_2(g) \rightarrow CO_2(g) + 2H_2O(\ell)$$

$$\Delta\nu(\text{gases}) = -0.5$$

$$\Delta_c H° = -726\ 500 - 0.5 \times 8.314 \times 298.15\ \text{J mol}^{-1}$$

$$= -727.7\ \text{kJ mol}^{-1}$$

(b) (1) $CH_3OH(\ell) + \tfrac{3}{2}O_2(g) \rightarrow CO_2(g) + 2H_2O(\ell)$

$$\Delta H° = 727.7\ \text{kJ mol}^{-1}$$

(2) $H_2(g) + \tfrac{1}{2}O_2(g) \rightarrow H_2O(\ell)$

$$\Delta_f H° = -285.85\ \text{kJ mol}^{-1}$$

(3) $C(s) + O_2(g) \rightarrow CO_2(g)$

$$\Delta_f H° = -393.51\ \text{kJ mol}^{-1}$$

2 x (2) + (3) – (1) gives

(4) $C(s) + 2H_2(g) + \tfrac{1}{2}O_2(g) \rightarrow CH_3OH(\ell)$

$$\Delta_f H° = 2(-285.85) - 393.51 + 727.7$$

$$= -237.5\ \text{kJ mol}^{-1}$$

(c) (5) $CH_3OH(\ell) \rightarrow CH_3OH(g) \quad \Delta_v H° = 35.27\ \text{kJ mol}^{-1}$

(4) + (5) gives

$$C(s) + 2H_2(g) + \tfrac{1}{2}O_2(g) \rightarrow CH_3OH(g)$$

$$\Delta_f H = -202.2\ \text{kJ mol}^{-1}$$

15. (1) $2C(graphite) + 3H_2(g) \rightarrow C_2H_6(g)$

$\Delta_f H° = -84.68 \text{ kJ mol}^{-1}$

(2) $C(graphite) + O_2(g) \rightarrow CO_2(g)$

$\Delta_f H° = -393.51 \text{ kJ mol}^{-1}$

(3) $H_2(g) + \frac{1}{2}O_2(g) \rightarrow H_2O(\ell)$

$\Delta_f H° = -285.85 \text{ kJ mol}^{-1}$

3 x (3) + 2 x (2) - (1) gives

$C_2H_6(g) + \frac{7}{2}O_2(g) \rightarrow 2CO_2(g) + 3H_2O(\ell)$

$\Delta_c H° = -1559.9 \text{ kJ mol}^{-1}$

16. Measure the heats of combustion of graphite, CO(g) and of $CO_2(g)$.

17. (1) $C_3H_6(g) + \frac{9}{2}O_2(g) \rightarrow 3CO_2(g) + 3H_2O(\ell)$

$\Delta_c H° = -2091.2 \text{ kJ mol}^{-1}$

(2) $C(graphite) + O_2(g) \rightarrow CO_2(g)$

$\Delta_f H° = -393.51 \text{ kJ mol}^{-1}$

(3) $H_2(g) + \frac{1}{2}O_2(g) \rightarrow H_2O(\ell)$

$\Delta_f H° = -285.85 \text{ kJ mol}^{-1}$

3 x (2) + 3 x (3) - (1) gives

$3C(graphite) + 3H_2(g) \rightarrow C_3H_6(g)$

$\Delta_f H° = 53.1 \text{ kJ mol}^{-1}$

18. $\Delta H° = -277.69 - 52.26 + 285.85$

$= -44.1 \text{ kJ mol}^{-1}$

19. (1) $C_2H_5OH + 3O_2 \rightarrow 2CO_2 + 3H_2O$

$\Delta_c H° = 1370.7 \text{ kJ mol}^{-1}$

(2) $CH_3CHO + \frac{5}{2}O_2 \rightarrow 2CO_2 + 2H_2O$

$$\Delta_c H^\circ = -1167.3 \text{ kJ mol}^{-1}$$

(3) $CH_3COOH + 2O_2 \rightarrow 2CO_2 + 2H_2O$

$$\Delta_c H^\circ = -876.1 \text{ kJ mol}^{-1}$$

Reaction (a) is (1) - (2); $\Delta H^\circ = -1370.7 + 1167.3$

$$= -203.4 \text{ kJ mol}^{-1}$$

Reaction (b) is (2) - (3); $\Delta H^\circ = -1167.3 + 876.1$

$$= -291.2 \text{ kJ mol}^{-1}$$

20. From Table 2.1

(1) $2C + 3H_2 + \frac{1}{2}O_2 \rightarrow C_2H_5OH$

$$\Delta_f H^\circ = -277.69 \text{ kJ mol}^{-1}$$

(2) $2C + 2H_2 + O_2 \rightarrow CH_3COOH$

$$\Delta_f H^\circ = -484.5 \text{ kJ mol}^{-1}$$

(3) $H_2 + \frac{1}{2}O_2 \rightarrow H_2O$

$$\Delta_f H^\circ = -285.85 \text{ kJ mol}^{-1}$$

Reaction is (2) + (3) - (1); $\Delta H^\circ = -492.7 \text{ kJ mol}^{-1}$

This differs slightly from the result of Problem 19, viz., $-203.4 - 291.2 = -494.6 \text{ kJ mol}^{-1}$

21. $\Delta H^\circ = 2(-1263.1) - (-2238.3) - (-285.85)$

$$= -2.05 \text{ kJ mol}^{-1}$$

22. $-131.4 \text{ kJ mol}^{-1} = \Delta_f H^\circ(\text{succinate}) - (-777.4 \text{ kJ mol}^{-1})$

$\Delta_f H^\circ = -908.8 \text{ kJ mol}^{-1}$

23. α-D-glucose (aq) → β-D-glucose (aq)

$$\Delta H^\circ = -1.16 \text{ kJ mol}^{-1}$$

α-D-glucose (s) → α-D-glucose (aq)

$$\Delta H^\circ = 10.72 \text{ kJ mol}^{-1}$$

β-D-glucose (aq) → β-D-glucose (s)

$$\Delta H^\circ = -4.68 \text{ kJ mol}^{-1}$$

Adding:

α-D-glucose (s) → β-D-glucose (s)

$$\Delta H^\circ = 4.88 \text{ kJ mol}^{-1}$$

24. From Table 2.2, $\Delta_f H^\circ$ at 25°C = −393.5 kJ mol^{-1}.

From the values in Table 2.1 we obtain

$$\Delta d = (44.22 - 29.96 - 16.86) = -2.60 \text{ J K}^{-1} \text{ mol}^{-1}$$

$$\Delta e = (8.79 - 4.77 - 4.18) \times 10^{-3}$$

$$= -0.16 \times 10^{-3} \text{ J K}^{-2} \text{ mol}^{-1}$$

$$\Delta f = -(8.62 - 8.54 - 1.67) \times 10^{5}$$

$$= 1.59 \times 10^{5} \text{ J K mol}^{-1}$$

Then, from Eq. 2.52,

$$\Delta H_{1000 \text{ K}}/\text{J mol}^{-1} = -393\ 500 - 2.60\ (1000 - 398) + \frac{1}{2}(-0.16 \times 10^{-3})(1000^2 - 298^2) - 1.59 \times 10^5 \left(\frac{1}{1000} - \frac{1}{298}\right)$$

$$= -393\ 500 - 1825 - 73 + 375$$

$$= -395\ 023$$

$$\Delta H_{100 \text{ K}} = -395.0 \text{ kJ mol}^{-1}$$

25. In propane there are 2 C–C bonds and 8 C–H bonds; heat of atomization is thus

$$(2 \times 348) + (8 \times 413) = 4000 \text{ kJ mol}^{-1}$$

Then

$2C(g) + 8H(g) \rightarrow C_3H_8(g) \quad \Delta H^\circ = -4000 \text{ kJ mol}^{-1}$

$3C(\text{graphite}) \rightarrow 3C(g) \quad \Delta H^\circ = 2150.1 \text{ kJ mol}^{-1}$

$4H_2(g) \rightarrow 8H(g) \quad \Delta H^\circ = 1744 \text{ kJ mol}^{-1}$

Adding

$3C(\text{graphite}) + 4H_2(g) \rightarrow C_3H_8(g) \quad \Delta H^\circ = -105.9 \text{ kJ mol}^{-1}$

26. (a) 1 mol in 22.4 dm^3 at 273 K exerts a pressure of 1 atm; ∴ 2 mol in 11.2 dm^3 exert a pressure of 4 atm

(b) $PV = 4 \times 11.2 = 44.8$ atm dm^3

1 atm dm^3 = 101.3 J

44.8 atm dm^3 = 4.54 kJ

(c) $C_{V,m} = C_{P,m} - R = 29.4 - 8.3 = 21.1 \text{ J K}^{-1} \text{ mol}^{-1}$

27. (a) Zero

(b) 2 x 21.1 x 100 = 4220 J = 4.22 kJ

(c) 4220 J = 4.22 kJ

(d) $P = 4 \times 373/273 = 5.47$ atm = 554.2 kPa

(e) $P_2V_2 = 5.47 \times 11.2 = 61.3$ atm dm^3

= 6210 J = 6.21 kJ

(f) $\Delta H = 4220 + 6210 - 4540 = 5890$ J = 5.89 kJ

28. (a) $V = 11.2 \times 373/273 = 15.3$ dm^3

(b) $w = -P\Delta V = -4 \times (15.3 - 11.2) = -16.4$ atm dm^3

$= -1660$ J = 1.66 kJ

(c) $q = 2 \times 29.4 \times 100 = 5880$ J = 5.88 kJ

(d) $\Delta H = 5880\ \text{J} = 5.88\ \text{kJ}$

(e) $\Delta U = \Delta H - P\Delta V = 5880 - 1660 = 4220\ \text{J} = 4.22\ \text{kJ}$

$(= 2C_{V,m}\Delta T = 21.1 \times 2 \times 100)$

29. (a) Zero

(b) $P_2 = 8\ \text{atm} = 810.6\ \text{kPa}$

(c) $w = 2 \times 8.314 \times 273 \ln 2 = 3150\ \text{J} = 3.15\ \text{kJ}$

(d) $-q = 3150\ \text{J} = 3.15\ \text{kJ}$

(e) Zero

30. Initial volume of gas $= \dfrac{nRT}{P} = \dfrac{2 \times 0.082\ 05 \times 273.15}{10}$

$= 4.482\ \text{dm}^3$

Final volume $= 22.41\ \text{dm}^3$

$\Delta V = 17.93\ \text{dm}^3$

(a) Work done by the gas $= 2.0 \times 17.93$

$= 35.86\ \text{atm dm}^3 = 3633\ \text{J}$ = energy transferred to surroundings

(b) $\Delta U = \Delta H = 0$

(c) $q = 3633\ \text{J}$

31. (a) Work done by gas $= nRT \ln \dfrac{V_{\text{final}}}{V_{\text{initial}}}$

$= 2 \times 8.314 \times 273.15 \ln 5$

$= 7310\ \text{J}$

(b) $\Delta U = \Delta H = 0$

(c) $q = 7310\ \text{J} = 7.31\ \text{kJ}$

32. $C_{V,m} = 28.80 - 8.314 = 20.49\ \text{J K}^{-1}\ \text{mol}^{-1}$

$\gamma = \dfrac{28.80}{20.49} = 1.406$

(a) $P_2 = P_1\left(\frac{V_1}{V_2}\right)^{\gamma} = 3.0\left(\frac{1.5}{5.0}\right)^{1.406}$

$= 0.552$ atm

$T_2 = T_1\left(\frac{V_1}{V_2}\right)^{\gamma-1} = 298.15\left(\frac{1.5}{5.0}\right)^{0.406}$

$= 182.9$ K

(b) $\Delta U_m = C_{V,m}(T_2 - T_1)$

$= 20.49\ (182.9 - 298.15) = -2361$ J mol^{-1}

Amount of H_2, $n = \dfrac{3.0 \times 1.5}{0.082\ 05 \times 298.15} = 0.184$ mol

$\Delta U = -434$ J for 0.184 mol

$\Delta H_m = 28.80\ (182.9 - 298.15) = -3319$ J mol^{-1}

$\Delta H = -610$ J for 0.184 mol

33. (a) Initial volume, $V_1 = \dfrac{nRT}{P} = 0.1 \times 0.082\ 05 \times 353.15 = 2.898$ dm^3

$V_2 = V_1\left(\frac{P_1}{P_2}\right)^{1/\gamma}$

$= 2.898\ (10)^{0.763} = 16.81$ dm^3

(b) $T_2 = \dfrac{T_1P_2V_2}{P_1V_1} = \dfrac{353.15 \times 16.81}{10 \times 2.898} = 204.8$ K

(c) $C_{P,m} - C_{V,m} = 8.314$ J K^{-1} mol^{-1}

$\dfrac{C_{P,m}}{C_{V,m}} = 1.31$

$\therefore\ 0.31\ C_{V,m} = 8.314$ J K^{-1} mol^{-1}

$C_{V,m} = 26.82$ J K^{-1} mol^{-1}

$C_{P,m} = 35.13$ J K^{-1} mol^{-1}

$\Delta U_m = 26.82\ (204.8 - 353.15) = -3979\ \text{J K}^{-1}\ \text{mol}^{-1}$

$\Delta U = -398\ \text{J K}^{-1}$ for 0.1 mol

$\Delta H_m = 35.13\ (204.8 - 353.15) = -5212\ \text{J K}^{-1}\ \text{mol}^{-1}$

$\Delta H = -521\ \text{J K}^{-1}$ for 0.1 mol

34. (a) $C_{P,m} - C_{V,m} = R = 8.314\ \text{J K}^{-1}\ \text{mol}^{-1}$

$C_{P,m} = 29.84 + 8.2 \times 10^{-3}\ (T/\text{K})$

(b) $\Delta T = 0$, $\Delta U = \Delta H = 0$. It could be made to occur adiabatically by allowing free expansion, with $w = q = 0$.

35. The accompanying diagram shows two adiabatics intersected by two isotherms corresponding to temperatures T_h and T_c (compare Figure 3.2):

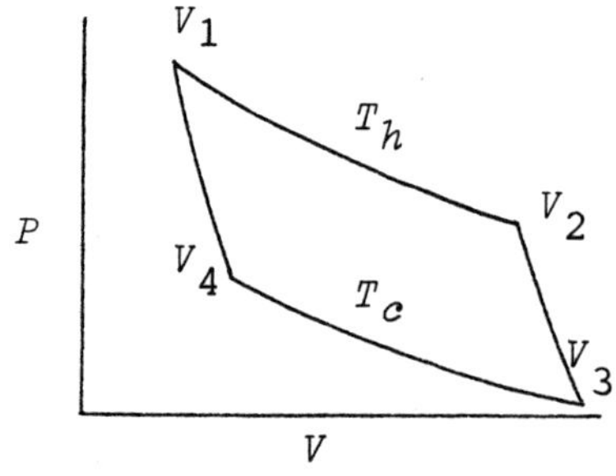

From Eq. 2.90

$T_h/T_c = (V_4/V_1)^{\gamma}$ and $T_h/T_c = (V_3/V_2)^{\gamma}$

Thus

$V_4/V_1 = V_3/V_2$ or $V_2/V_1 = V_3/V_4$

Thus if any isotherm is drawn to intersect the two adiabatics, the ratio of the volumes at the intersection points is always the same. The adiabatics therefore cannot intersect.

36. By definition

$$C_V = \left(\frac{\partial U}{\partial T}\right)_V$$

$$\left(\frac{\partial C_V}{\partial V}\right)_T = \frac{\partial}{\partial V}\left(\frac{\partial U}{\partial T}\right)_V = \frac{\partial}{\partial T}\left(\frac{\partial U}{\partial V}\right)_T$$

Since $(\partial U/\partial V)_T = 0$ for an ideal gas, $(\partial C_V/\partial V)_T = 0$, and C_V is therefore independent of V and of P.

Similarly,

$$C_P = \left(\frac{\partial H}{\partial T}\right)_P$$

and

$$\left(\frac{\partial C_P}{\partial P}\right)_T = \frac{\partial}{\partial P}\left(\frac{\partial H}{\partial T}\right)_P = \frac{\partial}{\partial T}\left(\frac{\partial H}{\partial P}\right)_T$$

Thus $(\partial C_p/\partial P)_T = 0$; C_P is therefore independent of P and of V.

37. The reversible work done by a gas in a reversible isothermal expansion is

$$-w = nRT \ln \frac{V_2}{V_1} \qquad \text{(Eq. 2.71)}$$

In this case $n = 1$ mol and $V_2/V_1 = 2$. Thus

$$1000 \text{ J mol}^{-1} = (8.314 \text{ J K}^{-1} \text{ mol}^{-1}) T \ln 2$$

$$T = \frac{1000}{8.314 \ln 2} \text{ K} = 173.5 \text{ K}$$

38. $C_{V,m} = \frac{5}{2}R$ and $C_{P,m} = \frac{7}{2}R$

$$\gamma = \frac{7}{5}$$

$$\frac{T_2}{T_1} = \left(\frac{V_1}{V_2}\right)^{\gamma-1} \qquad \text{(Eq. 2.90)}$$

and therefore

$$T_2 = T_1\left(\frac{V_1}{V_2}\right)^{\gamma-1} = 298.15 \text{ K x } (1/2)^{2/5} = 226.0 \text{ K}$$

$$\Delta U_m = C_{V,m}(T_2 - T_1) = \frac{5}{2}R(226.0 - 298.15) \text{ K}$$

$$= -1501 \text{ J mol}^{-1} = -1.5 \text{ kJ mol}^{-1}$$

$$\Delta H_m = C_{P,m}(T_2 - T_1) = \frac{7}{2}R(226.0 - 298.15) \text{ K}$$

$$= -2100 \text{ J mol}^{-1} = -2.1 \text{ kJ mol}^{-1}$$

39. (a) $\Delta H = C_P \Delta T$

$$C_{V,m} = \frac{3}{2}R = 12.47 \text{ J K}^{-1} \text{ mol}^{-1}$$

$$C_{P,m} = \frac{5}{2}R = 20.79 \text{ J K}^{-1} \text{ mol}^{-1}$$

$$1500 \text{ J mol}^{-1} = 20.79\ (T_2 - 300 \text{ K}) \text{ J K}^{-1} \text{ mol}^{-1}$$

$$T_2 = 372.15 \text{ K}$$

$$\frac{P_2}{P_1} = \frac{T_2}{T_1} \cdot \frac{V_1}{V_2} = \frac{372.15}{300} \cdot \frac{1}{2} = 0.62$$

$$P_2 = 0.62 \text{ atm}$$

(b) $\Delta U_m = C_{V,m}\Delta T$

$$= 12.47 \text{ x } 72.15 \text{ J mol}^{-1} = 900 \text{ J mol}^{-1}$$

$$w = \Delta U - q = (900 - 1000) \text{ J mol}^{-1} = -100 \text{ J mol}^{-1}$$

40. (a) Work done on the system $= -\int_{V_1}^{V_2} P dV = RT \ln \frac{V_1}{V_2}$

$$= 8.314 \text{ x } 300 \ln \frac{10}{0.2} = 9757 \text{ J} = 9.76 \text{ kJ mol}^{-1}$$

(b) $w = -RT \int_{V_1}^{V_2} \frac{dV}{V - b}$ Put $V - b = x;\ dV = dx$

$$= -RT \int_{V_1}^{V_2} \frac{dx}{x} = -RT \ln (V - b) \Big|_{V_1}^{V_2}$$

$$= RT \ln \frac{V_1 - b}{V_2 - b} = 8.314 \text{ x } 300 \ln \frac{9.97}{0.17}$$

$$= 10\ 155 \text{ J mol}^{-1} = 10.16 \text{ kJ mol}^{-1}$$

More work is done in (b) because of the greater ratio of free volumes.

41. (a) $w = RT \ln \frac{V_1}{V_2} = 8.314 \times 100 \ln \frac{20}{5}$

$= 1153 \text{ J mol}^{-1} = 1.15 \text{ kJ mol}^{-1}$

(b) $P = \frac{RT}{V} - \frac{a}{V^2}$

$$w = -\int_{V_1}^{V_2} PdV = -RT \int_{V_1}^{V_2} \frac{dV}{V} + a \int_{V_1}^{V_2} \frac{dV}{V^2}$$

$$= -RT \ln \frac{V_1}{V_2} - a\left(\frac{1}{V_2} - \frac{1}{V_1}\right)$$

$$= 1153 - 3.8 \times 101.3 \left(\frac{1}{5} - \frac{1}{20}\right)$$

$$= 1095 \text{ J} = 1.20 \text{ kJ mol}^{-1}$$

In (b), less work has to be done against the molecular repulsions.

42. For 1 mol of a van der Waals gas

$$\left(P + \frac{a}{V_m^2}\right)(V_m - b) = RT \qquad \text{(Eq. 1.88)}$$

Therefore

$$T = \frac{P(V_m - b)}{R} + \frac{a}{V_m^2}(V_m - b)$$

and

$$\left(\frac{\partial T}{\partial P}\right)_V = \left(\frac{V_m - b}{R}\right)$$

43. The definition of C_V is

$$C_V \equiv \left(\frac{\partial U}{\partial T}\right)_V$$

From Euler's chain rule (Appendix C)

$$\left(\frac{\partial U}{\partial T}\right)_V = -\left(\frac{\partial U}{\partial V}\right)_T \left(\frac{\partial V}{\partial T}\right)_V$$

Therefore

$$C_V = -\left(\frac{\partial U}{\partial V}\right)_T \left(\frac{\partial V}{\partial T}\right)_V$$

44. We have the general relationship (Appendix C)

$$dP = \left(\frac{\partial P}{\partial T}\right)_V dT + \left(\frac{\partial P}{\partial V}\right)_T dV$$

For an ideal gas

$$P = \frac{nRT}{V}$$

$$\left(\frac{\partial P}{\partial T}\right)_V = \frac{nR}{V} = \frac{P}{T}\ ; \left(\frac{\partial P}{\partial V}\right)_T = -\frac{nRT}{V^2} = -\frac{P}{V}$$

Therefore

$$dP = \frac{P}{T}\,dT - \frac{P}{V}\,dV$$

and

$$\frac{1}{P}\frac{dP}{dt} = \frac{1}{T}\frac{dT}{dt} - \frac{1}{V}\frac{dV}{dt}$$

45. We have the general relationship (Appendix C)

$$dP = \left(\frac{\partial P}{\partial V_m}\right)_T dV_m + \left(\frac{\partial P}{\partial T}\right)_{V_m} dT$$

For one mole of a van der Waals gas

$$P = \frac{RT}{V_m - b} - \frac{a}{V_m^2}$$

Then

$$\left(\frac{\partial P}{\partial T}\right)_{V_m} \quad \frac{R}{V_m - b}\ ; \left(\frac{\partial P}{\partial V_m}\right)_T = -\frac{RT}{(V_m - b)^2} + \frac{a}{V_m^3}$$

Therefore

$$dP = -\frac{RT\ dV_m}{(V_m - b)^2} + \frac{a\ dV_m}{V_m^3} + \frac{R\ dT}{V_m - b}$$

$$= -\frac{P + \dfrac{a}{V_m^2}}{V_m \quad b}\,dV_m + \frac{a\ dV_m}{V_m^3} + \frac{P\ dT}{T} - \frac{a}{V_m^2}\,\frac{dT}{T}$$

$$= -\frac{P}{V_m - b}\,dV_m - \frac{ab}{V_m^3(V_m - b)}\,dV_m + \frac{P\ dT}{T} - \frac{a\ dT}{V_m^2 T}$$

46. The reaction is

$CH_4(g) + 2O_2(g) \rightarrow CO_2(g) + 2H_2O(g)$

$\Delta H° = \Delta_f H°(CO_2,g) + 2\Delta_f H°(H_2O,g) - \Delta_f H°(CH_4,g)$

$\Delta H°/kJ\ mol^{-1} = -393.51\ (2 \times 241.84) + 74.81 = 802.38$

For the product gases, $CO_2 + 2H_2O$,

$C_p(CO_2)/J\ K^{-1}\ mol^{-1} = 44.22 + 8.79 \times 10^{-3}\ (T/K)$

$2C_p(H_2O)/J\ K^{-1}\ mol^{-1} = 61.08 + 2.06 \times 10^{-2}\ (T/K)$

$C_p(products)/J\ K^{-1} = 105.30 + 2.939 \times 10^{-2}\ (T/K)$

$$802\ 380 = \int_{298.15}^{T_2/K} (105.30 + 2.939 \times 10^{-2}(T/K)\,d(T/K)$$

$$= 105.30\ [T_2/K - 298.15]$$
$$+ (2.93 \times 10^{-2})[T_2/K)^2 - 298.15^2]$$

$$1.470 \times 10^{-2}\ (T_2/K)^2 + 105.30\ (T_2/K) - 835\ 081 = 0$$

$$T_2/K = \frac{-105.30 \pm \overline{11\ 088 + 49\ 103}}{2.94 \times 10^{-2}}$$

$$= 4763$$

$$T_2 = 4760\ K$$

47. The initial temperature is T_1 (= 298.15 K) and the final temperature T_2. Then

$$dU = nC_{V,m}\, dT$$

Also

$$dU - dq + dw = -P_2 dV \text{ since } dq = 0$$

Thus

$$nC_{V,m}\, dT = -P_2 dV$$

$$nC_{V,m}(T_2 - T_1) = -P_2(V_2 - V_1)$$

$$= -P_2\left(\frac{nRT_2}{P_2} - \frac{nRT_1}{P_1}\right)$$

$$= -nRT_2 + \frac{nRT_1P_2}{P_1}$$

$$T_2 = \frac{C_{V,m}T_1 + \dfrac{RT_1P_2}{P_1}}{C_{V,m} + R}$$

$$= \frac{(20.8 \times 298.15) + (8.314 \times 298.15 \times 0.1}{20.8 + 8.314}\ \text{K}$$

$$= 221\ \text{K}$$

$$\Delta U = 5 \times 20.8\ (221 - 298.15) = -3030\ \text{J}$$

$$\Delta H = 5 \times 29.1\ (221 - 298.15) = -11\ 200\ \text{J}$$

CHAPTER 3: WORKED SOLUTIONS
THE SECOND AND THIRD LAWS OF THERMODYNAMICS

1. (a) Efficiency $= \dfrac{T_h - T_c}{T_h} = \dfrac{1000 - 200}{1000} = 0.8 = 80\%$

(b) Heat rejected = 150 x 1000/200 = 30 kJ

(c) Entropy increase $= \dfrac{150\ 000}{1000} = 150\ \text{J K}^{-1}$

(d) Entropy decrease $= \dfrac{30\ 000}{200} = 150\ \text{J K}^{-1}$

(e) $\Delta S = 0$

(f) $\Delta S = 0$

(g) $\Delta S = 150\ \text{J K}^{-1}$; $\Delta H = 0$

$\Delta G = \Delta H - T\Delta S = 0 - 1000 \times 150 = -150\ 000\ \text{J}$

$= -150\ \text{kJ}$

(h) $\Delta H = 200(1000 - 200)$

$= 160\ \text{kJ}$

$\Delta S = 0$

$\Delta G = 160\ \text{kJ}$

2. Efficiency $= \dfrac{w}{q_h} = \dfrac{398 - 313}{398} = 0.214$

$\dfrac{1500}{0.214} = 7010$ J must be withdrawn.

3. (a)

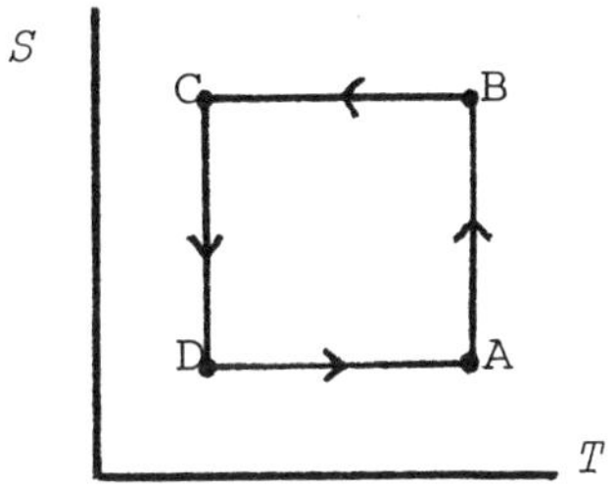

(b) $\dfrac{q_h}{T_h} + \dfrac{q_c}{T_c} = 0 = \dfrac{q_h}{T_h} + \dfrac{q_c}{300\ \text{K}} = 0$

Work performed by system, $-w = q_h + q_c = 10\ \text{kJ}$

$$\Delta S_h = \frac{q_h}{T_h} = 100\ \mathrm{J\ K^{-1}}$$

$$100\ \mathrm{J\ K^{-1}} + \frac{q_c}{300\ \mathrm{K}} = 0$$

$$q_c = -30\ 000\ \mathrm{J} = -30\ \mathrm{kJ}$$

$$q_h = 40\ \mathrm{kJ}$$

$$T_h = \frac{40}{30} \times 300\ \mathrm{K} = 400\ \mathrm{K}$$

4. C_6H_6: 30 800/353 = 87.3 J K^{-1} mol^{-1}

$CHCl_3$: 29 400/334 = 88.0 J K^{-1} mol^{-1}

H_2O: 40 600/373 = 108.8 J K^{-1} mol^{-1}

C_2H_5OH: 38 500/351 = 109.7 J K^{-1} mol^{-1}

} Hydrogen-bonded structure in liquids H_2O and C_2H_5OH.

5. (a) The reaction is

$$\mathrm{C(graphite)} + 2\mathrm{H_2(g)} + \tfrac{1}{2}\mathrm{O_2(g)} \rightarrow \mathrm{CH_3OH}(\ell)$$

$$\Delta_f H^\circ = 126.8 - [5.69 + (2 \times 130.5) + \tfrac{1}{2}(205.0)]$$

$$= -242.4\ \mathrm{J\ K^{-1}\ mol^{-1}}$$

(b) The reaction is

$$\mathrm{C(graphite)} + 2\mathrm{H_2(g)} + \tfrac{1}{2}\mathrm{O_2(g)} + \mathrm{N_2(g)} \rightarrow \mathrm{H_2NCONH_2(s)}$$

$$\Delta_f H^\circ = 104.6 - [5.69 + (2 \times 130.5) + \tfrac{1}{2}(205.0) + 191.6] = -456.2\ \mathrm{J\ K^{-1}\ mol^{-1}}$$

6. (a) $$\Delta S_m = \int_{298}^{353} \frac{C_{P,m}\,dT}{T} = \frac{5}{2} R \ln \frac{353}{298}$$

$$= 3.52\ \mathrm{J\ K^{-1}\ mol^{-1}}$$

(b) $$\Delta S_m = \int_{298}^{353} \frac{C_{V,m}\,dT}{T} = \frac{3}{2} R \ln \frac{353}{298}$$

$$= 2.11\ \mathrm{J\ K^{-1}\ mol^{-1}}$$

7. For N_2 and O_2, the volume increases by a factor of 2.5; increase of entropy for each

$\Delta S = R \ln 2.5 = 7.62 \text{ J K}^{-1}$

For H_2, volume increases by a factor of 5;

$\Delta S = 0.5 \times 8.314 \ln 5 = 6.69 \text{ J K}^{-1}$

Total $\Delta S = 2 \times 7.62 + 6.69 = 21.92 \text{ J K}^{-1}$

8. By Eq. 3.63

$$\Delta S = 8.314 \ln \frac{3}{1} + 5 \times 8.314 \ln \frac{3}{2}$$
$$= 9.13 + 16.86 = 25.99 \text{ J K}^{-1}$$

9. The mole fractions are

$x_{N_2} = 0.79 \qquad x_{O_2} = 0.20 \qquad x_{Ar} = 0.01$

The entropy change per mole of mixture is thus, by Eq. 3.65

$$\Delta S = -8.314(0.79 \ln 0.79 + 0.20 \ln 0.20 + 0.01 \ln 0.01)$$
$$= -8.314(-0.186 - 0.322 - 0.046) = 4.61 \text{ J K}^{-1} \text{ mol}^{-1}$$

10. The equation for the formation of ethanol from its elements is

$$2C(\text{graphite}) + 3H_2(g) + \frac{1}{2}O_2(g) \rightarrow C_2H_5OH(\ell)$$

The standard molar entropy of formation is thus

$$\Delta_f S_m^\circ = S^\circ_{\text{product}} - \Sigma S^\circ_{\text{reactants}}$$
$$= 160.7 - [(2 \times 5.69) + (3 \times 130.5) + \frac{1}{2}(205.0)]$$
$$= -344.7 \text{ J K}^{-1} \text{ mol}^{-1}$$

11. (a) $\Delta S_{\text{gas}} = R \ln \dfrac{V_2}{V_1} = 8.314 \ln 10 = 19.1 \text{ J K}^{-1} \text{ mol}^{-1}$

$\Delta S_{\text{surr}} = -19.1 \text{ J K}^{-1} \text{ mol}^{-1}$

(b) $\Delta U = q + w = 0$; since $q = 0$ and $w = 0$;

therefore

$\Delta S_{gas} = 19.1 \text{ J K}^{-1} \text{ mol}^{-1}$; $\Delta S_{surr} = 0$ since $q = 0$

Net $\Delta S = 19.1 \text{ J K}^{-1} \text{ mol}^{-1}$

12. (a) Positive (increase in number of molecules)

(b) Positive (decrease in electrostriction)

(c) Negative (increase in electrostriction)

(d) Positive (decrease in electrostriction)

13. Heat absorbed when temperature rises = $dq = C_P dT$

Corresponding entropy change, $\Delta S = \int \frac{C_P dT}{T}$

Entropy increase when the temperature rises from T_1 to T_2:

$$\int_{T_1}^{T_2} \frac{C_P}{T} dT = n \int_{T_1}^{T_2} \frac{C_{P,m}}{T} dT$$

If the gas is ideal C_P is constant and

$$\Delta S = C_P \ln \frac{T_2}{T_1} = nC_{P,m} \ln \frac{T_2}{T_1}$$

14. By equation 3.57 the entropy change is

$$\Delta S = 5 \times 12.5 \ln \frac{373}{300} + 5 \times 8.314 \ln \frac{10}{5}$$

$$= 13.61 + 28.1 = 42.4 \text{ J K}^{-1}$$

15. Let the value of the final Celsius temperature be T^u. Then $200 \times 0.140\ (100 - T^u) = 100 \times 4.18\ (T^u - 20)$

$$6.70 - 0.0670\ T^u = T^u - 20$$

$$T^u = \frac{26.70}{1.0650} = 25.02; \; T = 25.02°C$$

(a) $$\Delta S_{mercury} = 200 \times 0.140 \int_{373.15}^{298.1} \frac{dT}{T}$$

$$= 200 \times 0.140 \ln \frac{298.17}{373.15}$$

$$= -6.28 \text{ J K}^{-1}$$

(b) $$\Delta S_{\text{water + vessel}} = 100 \times 4.18 \int_{293.15}^{298.17} \frac{dT}{T}$$

$$= 100 \times 4.18 \ln \frac{298.17}{293.15}$$

$$= 7.10 \text{ J K}^{-1}$$

(c) Net $\Delta S = -6.28 + 7.10 = 0.82 \text{ J K}^{-1}$

16. Heat required to melt 20 g of ice

$$= \frac{20}{18} \times 6020 = 6689 \text{ J}$$

Heat required to heat 20 g of water from 0°C to T°C is

$$(20 \times 4.184\, T) \text{ J}$$

Heat required to cool 70 g of water from 30°C to T°C is

$$70 \times 4.184\, (30 - T) \text{ J}$$

Heat balance equation:

$$6689 + 20 \times 4.184\, T = 70 \times 4.184\, (30 - T)$$

$$T = 5.57°C$$

Reversible processes:

(a) Cool 70 g of water to 0°C:

$$\Delta S_{system} = 70 \times 4.184 \ln \frac{273.15}{303.15} = -30.52 \text{ J K}^{-1}$$

$$\Delta S_{surr} = 30.52 \text{ J K}^{-1}$$

(b) Melt 20 g of ice at 0°C:

$$\Delta S_{system} = \frac{20 \times 6020}{18 \times 273.15} = 24.58 \text{ J K}^{-1}$$

$$\Delta S_{surr} = -24.58 \text{ J K}^{-1}$$

(c) Heat 90 g of water to 5.57°C:

$$\Delta S_{system} = 90 \times 4.184 \ln \frac{278.72}{273.15} = 7.60 \text{ J K}^{-1}$$

$$\Delta S_{surr} = -7.60 \text{ J K}^{-1}$$

Net $\Delta S_{system} = 1.66 \text{ J K}^{-1}$; $\Delta S_{surr} = -1.66 \text{ J K}^{-1}$

If the process is carried out irreversibly

$\Delta S_{system} = 1.66 \text{ J K}^{-1}$; $\Delta S_{surr} = 0$

Net $\Delta S = 1.66 \text{ J K}^{-1}$

17. $$\Delta S_m = \int_{300}^{1000} \frac{C_{P,m} dT}{T}$$

$$\Delta S_m/\text{J K}^{-1} \text{ mol}^{-1} = \int_{300}^{1000} \frac{28.58}{T/\text{K}} + 0.00376 - \frac{50\ 000}{(T/\text{K})^3}\, d(T/\text{K})$$

$$= 28.58 \ln \frac{1000}{300} + 0.00376\ (1000 - 300) + 25\ 000 \left(\frac{1}{1000^2} - \frac{1}{300^2}\right)$$

$$= 34.41 + 2.63 - 0.25 = 36.79$$

$$\Delta S_m = 36.8 \text{ J K}^{-1} \text{ mol}^{-1}$$

18. The work done by the system in an isothermal reversible expansion is

$$-w_{rev} = nRT \ln \frac{V_2}{V_1}$$

The entropy of expansion is

$$\Delta S = nR \ln \frac{V_2}{V_1}$$

If the process were reversible, the work done by the

system would be

$$-w_{rev} = T\Delta S = 300 \times 50 = 15\ 000 \text{ J} = 15 \text{ kJ}$$

Since the actual work was less than this, the process is irreversible. Degree of irreversibility = $\frac{6}{15}$ = 0.4

19. From Eq. 3.91

$$\Delta G^\circ = 2(-181.75) + 2(-394.34) - (-914.54)$$

$$= -237.6 \text{ kJ mol}^{-1}$$

20. Work done by system = 30.19 dm^3 atm mol^{-1}

$$= 3060 \text{ J mol}^{-1}$$

$$\Delta H = 40\ 600 \text{ J mol}^{-1}$$

$$\Delta U = \Delta H - \Delta(PV)$$

$$= 40\ 600 - 3\ 060 = 37\ 540 \text{ J mol}^{-1}$$

$\Delta G = 0$ $\quad\quad$ $\Delta S = \frac{40\ 640}{373.15} = 108.8 \text{ J K}^{-1} \text{ mol}^{-1}$

21. (a) At 0°C, $\Delta H = -6020 \text{ J mol}^{-1}$

$$\Delta S = -22.0 \text{ J K}^{-1} \text{ mol}^{-1}$$

$$\Delta G = -6020 + 22.0 \times 273 = 0$$

(b) At -10°C, $\Delta H = -5644 \text{ J mol}^{-1}$

$$\Delta S = -20.65 \text{ J K}^{-1} \text{ mol}^{-1}$$

$$\Delta G = -5644 + 20.65 \times 263 = -213 \text{ J mol}^{-1}$$

22. $\Delta U = 0$

$$\Delta H = \Delta U + \Delta(PV) = 0$$

$$\Delta S = \frac{q_{rev}}{T} = \frac{1}{T}\int_{V_1}^{V_2} PdV = R \ln \frac{V_2}{V_1} = 8.314 \ln 10$$

$$= 19.14 \text{ J K}^{-1} \text{ mol}^{-1}$$

$\Delta A = \Delta U + T\Delta S = -298.15 \times 19.14 = -5.706$ kJ mol^{-1}

$\Delta G = -5.706$ kJ mol^{-1}

The quantities are all state functions, and the above values therefore do not depend on how the process is carried out.

23. (a) $\Delta G = -85\ 200 + (300 \times 170.2)$ J mol^{-1}

$= -34.14$ kJ mol^{-1}

(b) $\Delta G = -85\ 200 + (600 \times 170.2)$ J mol^{-1}

$= 16.92$ kJ mol^{-1}

(c) $\Delta G = -85\ 200 + (1000 \times 170.2)$ J mol^{-1}

$= 85.00$ kJ mol^{-1}

(d) At $T = 85\ 200/170.2 = 500.6$ K

24. Conversion of water to vapor at 0.0313 atm:

$\Delta H = 44.01$ kJ mol^{-1}

$$\Delta S = \frac{44\ 010}{298.15} = 147.6 \text{ J K}^{-1} \text{ mol}^{-1}$$

Reversible expansion from 0.0313 atm to 10^{-5} atm

$\Delta H = 0$

$$\Delta S = 8.314 \ln \frac{0.0313}{10^{-5}} = 66.9 \text{ J K}^{-1} \text{ mol}^{-1}$$

Net $\Delta H = 44.01$ kJ mol^{-1}

$\Delta S = 214.5$ J K^{-1} mol^{-1}

$\Delta G = 44\ 010 - 298.15 \times 214.5 = -19\ 940$ J mol^{-1}

$= -19.94$ kJ mol^{-1}

25. (a) ΔU, ΔH (c) ΔH

(b) ΔS (d) ΔH

(e) None (g) ΔU

(f) ΔG (h) None

26. $$\left(\frac{\partial G}{\partial P}\right)_T = V$$

and therefore

$$\Delta G = \int V dP$$

The molar volume of mercury is

$$V_m = \frac{200.6 \text{ g mol}^{-1}}{13.5 \text{ g cm}^{-3}} = 1.486 \times 10^{-5} \text{ m}^3 \text{ mol}^{-1}$$

Then $\Delta G_m = 1.486 \times 10^{-5} \text{ m}^3 \times 999 \text{ atm} \times 1.013\ 25 \times 10^5 \text{ Pa atm}^{-1}$

$$= 1504 \text{ J mol}^{-1} = 1.504 \text{ kJ mol}^{-1}$$

27. $$\left(\frac{\partial G_m}{\partial T}\right)_P = -S_m$$

$$= -(A + B \ln T)$$

where $A = 36.36 \text{ J K}^{-1} \text{ mol}^{-1}$

$B = 20.79 \text{ J K}^{-1} \text{ mol}^{-1}$

$$\Delta G_m = -\int_{298.15}^{323.15} (A + B \ln T)\, dT$$

$$= -[AT + B(T \ln T - T)] \Big|_{298.15}^{323.15}$$

$$= -(A - B)\, 25 - B\,(323.15 \ln 323.15 - 298.15 \ln 298.15)$$

$$= -389.3 - 3502.3 \text{ J mol}^{-1} = -3.89 \text{ kJ mol}^{-1}$$

28. ΔU and ΔH are zero for the isothermal expansion of an ideal gas.

$$q = -w = 500 \text{ J mol}^{-1}$$

$$\Delta S = R \ln 2 = 5.76 \text{ J K}^{-1} \text{ mol}^{-1}$$

$$\Delta G = -T\Delta S = -1729 \text{ J mol}^{-1} = -1.73 \text{ kJ mol}^{-1}$$

$$w_{rev} = 1.73 \text{ kJ mol}^{-1}; \quad q_{rev} = -1.73 \text{ kJ mol}^{-1}$$

29. No work is done; $w = 0$

$$\Delta U = q + w = q = 30 \text{ kJ mol}^{-1}$$

$$\Delta H = \Delta U + \Delta(PV) = \Delta U + RT$$

$$= 30\ 000 + 3102 = 33\ 102 \text{ J mol}^{-1} = 33.1 \text{ kJ mol}^{-1}$$

To obtain the entropy change, consider the reversible processes:

1) H_2O (ℓ, 100°C) → H_2O (g, 100°C, 1 atm)

$$\Delta S_1 = \frac{40\ 600}{373.15} = 108.8 \text{ J K}^{-1}$$

2) H_2O (g, 100°C, 1 atm) → H_2O (g, 100°C, 0.5 atm)

$$\Delta S_2 = R \ln \frac{V_2}{V_1} = R \ln 2 = 5.76 \text{ J K}^{-1} \text{ mol}^{-1}$$

Net $\Delta S = 114.6 \text{ J K}^{-1} \text{ mol}^{-1}$

Net $\Delta G = \Delta H - T\Delta S$

$$= 33\ 102 - 42\ 760 = -9660 \text{ J mol}^{-1}$$

$$= -9.66 \text{ kJ mol}^{-1}$$

30. Allow the process to occur by the following reversible steps:

1) H_2O (g, 100°C, 2 atm) → H_2O (g, 100°C, 1 atm)

$\Delta H_1 = 0$

$\Delta S_1 = R \ln V_2/V_1 = R \ln 2 = 5.76 \text{ J K}^{-1} \text{ mol}^{-1}$

$\Delta G_1 = -T\Delta S_1 = -2150 \text{ J mol}^{-1}$

2) H_2O (g, 100°C, 1 atm) → H_2O (ℓ, 100°C, 1 atm)

$\Delta H_2 = -40\ 600 \text{ J mol}^{-1}$

$\Delta S_2 = -\frac{40\ 600}{373.15} = -108.8 \text{ J K}^{-1} \text{ mol}^{-1}$

$\Delta G_2 = 0$ (reversible process at constant P and T)

3) H_2O (ℓ, 100°C, 1 atm) → H_2O (ℓ, 100°C, 2 atm)

The ΔH, ΔS, and ΔG changes are negligible for this process.

The overall changes are thus

$\Delta H = -40\ 600 \text{ J mol}^{-1} = -40.6 \text{ kJ mol}^{-1}$

$\Delta S = 5.76 - 108.8 = -103 \text{ J K}^{-1} \text{ mol}^{-1}$

$\Delta G = -2.15 \text{ kJ mol}^{-1}$

31. $\Delta U = \int_{T_1}^{T_2} C_{V,m} dT$

$$C_{V,m} = C_{P,m} - R$$
$$= (20.27 + 1.76 \times 10^{-2}\ T/\mathrm{K})\ \mathrm{J\ K^{-1}\ mol^{-1}}$$

$$q = 0;\ C_{V,m}dT = dw = -P_2 dV$$

$$\Delta U_m = \int_{300\ \mathrm{K}}^{T_2} (20.27 + 1.76 \times 10^{-2}\ T/\mathrm{K})\,dT$$
$$= -P_2 \int_{V_1}^{V_2} dV = -P_2\left(\frac{RT_2}{P_2} - \frac{300R}{P_1}\right)$$
$$= -4 \times 8.314\left(\frac{T_2}{4} - \frac{300}{10}\right)$$

$$20.27\ (T_2 - 300) + 0.0088\ (T_2{}^2 - 300^2)$$
$$= -4 \times 8.314\left(\frac{T_2}{4} - 30\right)$$

$$0.0088\ T_2{}^2 + 28.58\ T_2 - 7871 = 0$$

$$T_2 = \frac{-28.58 \pm (816.816 + 277.06)^{1/2}}{0.0176}$$
$$= \frac{-28.58 + 33.07}{0.0176} = 255.3\ \mathrm{K}$$

$$\Delta U_m = 20.27\ (255.3 - 300) + 0.0088\ (255.3^2 - 300^2)$$
$$= -1124.5\ \mathrm{J\ mol^{-1}}$$

$$\Delta H_m = 28.58\ (255.3 - 300) + 0.0088\ (255.3^2 - 300^2)$$
$$= -1495.9\ \mathrm{J\ mol^{-1}}$$

$$dS = \frac{dq}{T} = \frac{dU + PdV}{T}$$

For 1 mol of ideal gas, $PV_m = RT$

$$d(PV_m) = RdT = PdV_m + V_m dP$$

$$PdV_m = RdT - V_m dP = RdT - \frac{RT}{P}\,dP$$

$$dS_m = \frac{dU_m + RdT - \frac{RTdP}{P}}{T} = \frac{C_{P,m}dT}{T} - R\,\frac{dP}{P}$$

$$\Delta S_m/\text{J K}^{-1}\text{ mol}^{-1} = \int_{300}^{255.3} \frac{(28.58 + 0.0176\,T)\,dT}{T} - 8.314 \ln \frac{4}{10}$$

$$= 28.58 \ln \frac{255.3}{300} + 0.0176\,(255.3 - 300) + 7.618$$

$$= 2.22$$

$\Delta S_m = 2.22$ J K^{-1} mol^{-1}

32. 100 atm. engine: $T_h = 585$ K; $T_c = 303$ K

$$\text{Efficiency} = \frac{585 - 303}{585} = 48.2\%$$

5 atm. engine: $T_h = 425$; $T_c = 303$ K

$$\text{Efficiency} = \frac{425 - 303}{425} = 28.7\%$$

33. $\frac{w}{q_c} = \frac{293 - 269}{269} = 0.089$

$w = 0.089 \times 10^4 = 890 \text{ J min}^{-1} = 14.8 \text{ J s}^{-1}$

At 40% efficiency, power = $14.8/0.4 = 37.1 \text{ J s}^{-1}$

$= 37.1$ W.

34. Performance factor $= \frac{T_h}{T_h - T_c}$

(a) 59.6%; (b) 11.9%; (c) 6.6%

35. Efficiency $= \frac{1200}{2273} = 0.528$

1 liter = 8.0 kg $= \frac{8000 \text{ g}}{114.2 \text{ g mol}^{-1}} = 70.05$ mol.

Energy produced = $70.05 \text{ mol} \times 5500 \text{ kJ mol}^{-1}$

$= 3.85 \times 10^5$ kJ

Work = $0.528 \times 3.85 \times 10^5 = 2.03 \times 10^5$ kJ

36. Let q_h be the heat supplied to the building at 20°C and q_c be the heat taken in by the pump at 10°C:

$$\frac{q_h}{q_c} = \frac{293.15}{283.15}$$

Work supplied to heat pump

$$w = q_h - q_c = q_h\left(1 - \frac{283.15}{293.15}\right) = 0.034\ q_h$$

Let q_h' be the heat produced by the fuel at 1000°C and q_c' be the heat rejected at 20°C. Work performed by the heat engine and supplied to the heat pump is

$$w = q_h' - q_c' = q_h'\left(1 - \frac{293.15}{1273.15}\right) = 0.770\ q_h'$$

Thus

$$0.034\ q_h = 0.770\ q_h'$$

and the performance factor is

$$\frac{q_h}{q_h'} = \frac{0.770}{0.034} = 22.6$$

37. The heat that must be removed from 1 kg (= 55.5 mol) of water in order to freeze it is

$$6.02 \times 55.5 = 334 \text{ kJ} = q_c$$

This is the heat gained by the refrigerator.

If the efficiency were 100%

$$\frac{T_h}{T_c} = \frac{298.15}{273.15} = -\frac{q_h}{q_c} = \frac{-q_h}{334 \text{ kJ}}$$

Thus the heat discharged at 25 °C is

$$-q_h = 365 \text{ kJ}$$

Work required to be supplied to the refrigerator = 365 - 334 = 31 kJ.

With an efficiency of 40% the actual work will be

$$\frac{100}{40} \times 31 = 78 \text{ kJ}$$

and the heat discharged at 25 °C will be

$$334 + 78 = 412 \text{ kJ}$$

38. From the First and Second Laws

$dU = TdS - PdV$

and therefore

$$\left(\frac{\partial U}{\partial V}\right)_T = T\left(\frac{\partial S}{\partial V}\right)_T - P$$

$$= T\left(\frac{\partial P}{\partial T}\right)_V - P$$

using the Maxwell equation 3.124. From the van der Waals equation

$$\left(\frac{\partial P}{\partial T}\right)_V = \frac{R}{V_m - b} = \frac{1}{T}\left(P + \frac{a}{V_m^2}\right)$$

$$\left(\frac{\partial U}{\partial V}\right)_T = \frac{a}{V_m^2}$$

39. $$\mu \equiv \left(\frac{\partial T}{\partial P}\right)_H = \frac{-(\partial H/\partial P)_T}{(\partial H/\partial T)_P} = -\frac{1}{C_P}\left(\frac{\partial H}{\partial P}\right)_T$$

Since $dH = TdS + VdP$, $\left(\frac{\partial H}{\partial P}\right)_T = T\left(\frac{\partial S}{\partial P}\right)_T + V$

$$= -T\left(\frac{\partial V}{\partial T}\right)_P + V$$

from the Maxwell equation 3.125. Thus

$$\mu = \frac{T\left(\frac{\partial V}{\partial T}\right)_P - V}{C_P}$$

$$= \frac{T\left(\frac{\partial V_m}{\partial T}\right)_P - V_m}{C_{P,m}} \text{ for 1 mol.}$$

The equation $P(V_m - b) = RT$ applies to 1 mol of gas, and it follows that

$$\left(\frac{\partial V}{\partial T}\right)_P = \frac{R}{P}$$

whence

$$\mu = \frac{\frac{RT}{P} - V_m}{C_{P,m}}$$

40. (a) $\left(\frac{\partial G}{\partial T}\right)_P = -S$ (Eq. 3.119)

and therefore

$$\left(\frac{\partial^2 G}{\partial T^2}\right)_P = -\left(\frac{\partial S}{\partial T}\right)_P$$

$$dS = \frac{q_{\text{rev}}}{T} = \frac{C_P dT}{T}$$

$$\left(\frac{\partial S}{\partial T}\right)_P = \frac{C_P}{T}$$

$$C_P = -T\left(\frac{\partial^2 G}{\partial T^2}\right)_P$$

(b) $$\left(\frac{\partial C_P}{\partial P}\right)_T = \left[\frac{\partial}{\partial P}\left(\frac{\partial H}{\partial T}\right)_P\right]_T = \left[\frac{\partial}{\partial T}\left(\frac{\partial H}{\partial P}\right)_T\right]_P$$

$$\left(\frac{\partial H}{\partial P}\right)_T = -T\left(\frac{\partial V}{\partial T}\right)_P + V \text{ (see Problem 39)}$$

$$\left[\frac{\partial}{\partial T}\left(\frac{\partial H}{\partial P}\right)_T\right]_P = -T\left(\frac{\partial^2 V}{\partial T^2}\right)_P - \left(\frac{\partial V}{\partial T}\right)_P + \left(\frac{\partial V}{\partial T}\right)_P$$

$$\left(\frac{\partial C_P}{\partial P}\right)_T = -T\left(\frac{\partial^2 V}{\partial T^2}\right)_P$$

41. $$A = U - TS$$

$$dA = dU - TdS - SdT$$

At constant temperature

$$dA = dU - TdS = dq + dw - TdS$$

But $dq = TdS$ and therefore $dA = dw$.

42. $$dU = TdS - PdV \qquad \text{(Eq. 3.105)}$$

Dividing by dV at constant T:

$$\left(\frac{\partial U}{\partial V}\right)_T = T\left(\frac{\partial S}{\partial V}\right)_T - P = 0$$

But

$$\left(\frac{\partial S}{\partial V}\right)_T = \left(\frac{\partial P}{\partial T}\right)_V \qquad \text{(Eq. 3.124)}$$

and therefore

$$\left(\frac{\partial P}{\partial T}\right)_V = \frac{P}{T}$$

Integrating, $\ln P = \ln T + \text{const.}$

or $P \propto T$

Thus

$$PV = \text{const} \times T$$

43. By Euler's reciprocity theorem (Appendix C)

$$\left(\frac{\partial S}{\partial V}\right)_U = -\frac{(\partial U/\partial V)_S}{(\partial U/\partial S)_V}$$

$$\left(\frac{\partial U}{\partial V}\right)_S = -P \text{ (Eq. 3.116)} \qquad \left(\frac{\partial U}{\partial S}\right)_V = T \text{ (Eq. 3.116)}$$

$$\therefore \quad \left(\frac{\partial S}{\partial V}\right)_U = \frac{P}{T}$$

For an ideal gas U depends only on T so that

$$\left(\frac{\partial S}{\partial V}\right)_U = \left(\frac{\partial S}{\partial V}\right)_T$$

For an isothermal process involving n mol of an ideal gas

$$dS = nR\ d\ \ln V = \frac{nRdV}{V} = \frac{PdV}{T}$$

Thus $(\partial S/\partial V)_T = P/T$ and therefore $(\partial S/\partial V)_U = P/T$.

CHAPTER 4: WORKED SOLUTIONS
CHEMICAL EQUILIBRIUM

1.

	$2A$	$\rightleftharpoons$ Y	+ $2Z$
Initial amounts:	4	0	0 mol
Amounts at equilibrium:	1	1.5	3.0 mol
Concentrations at equilibrium:	$\frac{1}{5}$	$\frac{1.5}{5}$	$\frac{3.0}{5}$ mol dm^{-3}

$$K_c = \frac{(1.5/5)(3.0/5)^2}{(1.5)^2} \text{ mol dm}^{-3} = \frac{1.5 \times (3.0)^2}{5} \text{mol dm}^{-3}$$

$$= 2.7 \text{ mol dm}^{-3}$$

2.

	A +	B $\rightleftharpoons$	Y +	Z
Initial amounts:	x	3	0	0 mol
Amounts at equilibrium:	$x - 2$	1	2	2 mol

$$\frac{2 \times 2}{x - 2} = 0.1; \ x - 2 = 40$$

$$x = 42$$

Thus initially there must be 42 mol of A.

3.

	A +	$2B$ $\rightleftharpoons$	Z
Initial amounts:	x	4	0 mol
Amounts at equilibrium:	$x - 1$	2	1 mol
Concentrations at equilibrium:	$\frac{x-1}{5}$	$\frac{2}{5}$	$\frac{1}{5}$ mol dm^{-3}

$$\frac{1/5}{[(x - 1)/5](2/5)^2} = 0.25 = \frac{25}{4(x - 1)}$$

$$25 = x - 1 \qquad x = 26$$

Thus initially there must be 26 mol of A.

4. Two moles of $SO_3(g)$ produce 3 mol of product; thus $\Sigma\nu = +1$ mol. Then, from Eq. 4.27,

$$K_p = K_c(RT)^{+1}$$

$$= (0.0271 \text{ mol dm}^{-3})(8.314 \text{ J K}^{-1} \text{ mol}^{-1} \text{ 1100 K})$$

$= 247.8 \text{ J dm}^{-3} = 2.478 \times 10^{5} \text{ J m}^{-3}$

$= 2.478 \times 10^{5} \text{ Pa} = 2.45 \text{ atm}$

5. $I_2 \rightleftharpoons 2I$

	I_2	$2I$
Initial amounts:	0.0061	0 mol
Equilibrium amounts:	0.0061(1 - 0.0274)	0.0061 x 2 x 0.0274 mol
	$= 5.93 \times 10^{-3}$	$= 3.3428 \times 10^{-4}$ mol

$$K_c = \frac{(3.3428 \times 10^{-4}/0.5)^2}{5.93 \times 10^{-3}/0.5} \text{ mol dm}^{-3} = 3.77 \times 10^{-5} \text{ mol dm}^{-3}$$

$$K_P = 3.77 \times 10^{-2} \text{ mol dm}^{-3}(8.314 \text{ J K}^{-1} \text{ mol}^{-1} \times 900 \text{ K})$$

$$= 2.82 \text{ Pa} = 2.78 \times 10^{-3} \text{ atm}$$

6. Addition of N_2 at constant *volume* and temperature necessarily requires the equilibrium to shift to the *right*.

$$K_c = \frac{[NH_3]^2}{[N_2][H_2]^3} = \frac{n^2_{NH_3} V}{n_{N_2} n^3_{H_2}}$$

If n_{N_2} is increased at constant V, the equilibrium must shift so as to produce more ammonia. If the *pressure* (as well as the temperature) is held constant, however, addition of N_2 requires that V is increased. If the increase in V^2 is greater than the increase in n_{N_2} the equilibrium will shift to the *left* when N_2 is added.

The volume V is proportional to $n_{N_2} + n_{H_2} + n_{NH_3}$, and V^2 is proportional to

$$(n_{N_2} + n_{H_2} + n_{NH_3})^2.$$

If n_{N_2} is very much larger than $n_{H_2} + n_{NH_3}$, V^2 will increase approximately with $n^2_{N_2}$, and therefore increases more strongly than n_{N_2}. If n_{N_2} is not much larger than $n_{H_2} + n_{NH_3}$, an increase in n_{N_2} will have a relatively smaller effect on V^2. The increase in ammonia dissociation when N_2 is added is therefore expected when N_2 is in excess, but not otherwise.

On the other hand, $n^3_{H_2}$ appears in the equilibrium expression; this varies more strongly than V^2, and added H_2 therefore cannot lead to the dissociation of ammonia.

7. Suppose that in 1 dm^3 there are

$$x \text{ mol of } N_2O_4$$

$$y \text{ mol of } NO_2$$

Pressure = 0.589 atm

$$= \frac{(x + y)\text{mol} \times 0.082\ 05 \text{ atm dm}^3 \text{ K}^{-1} \text{ mol}^{-1} \times 298.15 \text{ K}}{1 \text{ dm}^3}$$

$$\therefore x + y = 0.024\ 08$$

$$\text{Density} = 1.477 \text{ g dm}^{-3} = \frac{(92.02x + 46.01y)\text{g}}{1 \text{ dm}^3}$$

$2x + y = 0.032\ 10$; $x = 0.008\ 025$; $y = 0.016\ 05$

The amounts of N_2O_4 and NO_2 are in the ratio 1/2, and the degree of dissociation of N_2O_4 is thus 0.5:

$$K_c = \frac{(0.016\ 05)^2}{0.008\ 02} \text{ mol dm}^{-3} = 0.0321 \text{ mol dm}^{-3}$$

$K_P = K_c RT$ (from Eq. 4.26, with $\Sigma\nu = 1$)

$= 0.0321 \text{ mol dm}^{-3} \times 0.082\ 05 \times 298.15 \text{ dm}^3 \text{ atm mol}^{-1}$

$= 0.787$ atm

$K_x = K_P P^{-1}$ (from Eq. 4.32)

$= K_P\ (1 \text{ atm})^{-1} = 0.787$

Addition of He produces no effect, since concentrations, partial pressures and mole fractions remain unchanged.

8. $$1.10 \text{ g NOBr} = \frac{1.10}{14.01 + 16.00 + 79.91} = 0.010 \text{ mol}$$

If α is the degree of dissociation,

$n_{NOBr} = 0.01\ (1 - \alpha)$ mol; $n_{NO} = 0.01\ \alpha$ mol

$n_{Br_2} = 0.005\ \alpha$ mol

Total amount, $n = 0.01 + 0.005\ \alpha$ mol

$$= \frac{PV}{RT} = \frac{0.35 \text{ atm} \times 1 \text{ dm}^3}{0.082\ 05 \times 298.15 \text{ atm dm}^3 \text{ mol}^{-1}}$$

$= 0.0143$ mol

$\alpha = 0.861$; $n = 0.0143$

$n_{NOBr} = 1.39 \times 10^{-3}$ mol; $n_{NO} = 8.61 \times 10^{-3}$ mol

$n_{Br_2} = 4.305 \times 10^{-3}$ mol

$$K_c = \frac{(8.61 \times 10^{-3} \text{ mol dm}^{-3})^2\ (4.305 \times 10^{-3} \text{ mol dm}^{-3})}{(1.39 \times 10^{-3} \text{ mol dm}^{-3})^2}$$

$= 0.165 \text{ mol dm}^{-3}$

$K_P = K_cRT$ (from Eq. 4.26)

$= 0.165 \text{ mol dm}^{-3} \times 0.082\ 05 \times 298.15 \text{ dm}^3 \text{ atm mol}^{-1}$

$= 4.04$ atm

$K_x = K_P\ (0.35 \text{ atm})^{-1}$ (from Eq. 4.32)

$= 11.5$

9. Suppose that, if there were no dissociation, the partial pressure of $COCl_2$ was P; then the actual partial pressures are

$$COCl_2(g) \rightleftharpoons CO(g) + Cl_2(g)$$
$$P(1-\alpha) \qquad P\alpha \qquad P\alpha$$

$P + P\alpha = 2$ atm

With $\alpha = 6.30 \times 10^{-5}$,

$$P = \frac{2 \text{ atm}}{1 + 6.3 \times 10^{-5}} \approx 2 \text{ atm}$$

$$K_P = \frac{(2 \times 6.3 \times 10^{-5})^2}{2(1 - 6.3 \times 10^{-5})} \approx 2 \times (6.3 \times 10^{-5})^2 \text{ atm}$$

$= 7.94 \times 10^{-9}$ atm

$K_c = K_P(RT)^{-1}$ (from Eq. 4.26)

$= 7.94 \times 10^{-9}$ atm $\times$

$(0.082\ 05 \times 373.15 \text{ dm}^3 \text{ atm mol}^{-1})^{-1}$

$= 2.59 \times 10^{-10} \text{ mol dm}^{-3}$

$K_x = K_PP^{-1}$ (from Eq. 4.32)

$= 7.94 \times 10^{-9} \text{ atm } (2 \text{ atm})^{-1} = 3.97 \times 10^{-9}$

10. $H_2 + I_2 \rightleftharpoons 2HI$

	H_2	I_2	$2HI$
Initially:	1	3	0 mol
At equilibrium:	$1 - \frac{x}{2}$	$3 - \frac{x}{2}$	mol
After addition of 2 mol H_2:	$3 - x$	$3 - x$	$2x$ mol

$$K = \frac{x^2}{(1 - \frac{x}{2})(3 - \frac{x}{2})} = \frac{4x^2}{(3 - x)^2}$$

$x = 3/2$

$$K = \frac{4(3/2)^2}{(3/2)^2} = 4$$

11. 12.7 g iodine = 0.05 mol I_2

When all of the solid iodine has just gone the iodine pressure is 0.10 atm. The consumption of 0.05 mol I_2 leads to the formation of 0.10 mol HI, which exerts a pressure of

$$P_{HI} = \frac{0.1 \text{ mol} \times 0.082\ 05 \text{ dm}^3 \text{ atm K}^{-1} \text{ mol}^{-1} \times 313.15 \text{ K}}{10 \text{ dm}^3}$$

$$= 0.257 \text{ atm}$$

Then, if P_{H_2} is the partial pressure of H_2 after equilibrium is established,

$$\frac{(0.257 \text{ atm})^2}{P_{H_2} \times 0.1 \text{ atm}} = 20$$

$$P_{H_2} = 0.033 \text{ atm}$$

$$n_{H_2} = \frac{0.033 \text{ atm} \times 10 \text{ dm}^3}{0.082\ 05 \times 313.15 \text{ atm dm}^3 \text{ mol}^{-1}}$$

$$= 0.013 \text{ mol}$$

Thus 0.013 mol of H_2 are produced in the equilibrium mixture, and 0.05 mol of H_2 are required to remove the 0.05 mol of I_2. Therefore 0.063 mol of H_2 must be added.

12. (a) $\Delta G^\circ/\text{J mol}^{-1} = -8.314 \times 283.15 \ln (2.19 \times 10^{-3})$

$\Delta G^\circ = 14\ 416 \text{ J mol}^{-1} = 14.42 \text{ kJ mol}^{-1}$

(b) $(C_6H_5COOH)_2 \rightleftharpoons 2C_6H_5COOH$

	$(C_6H_5COOH)_2$	$2C_6H_5COOH$
Initially	0	0.1 mol dm^{-3}
At equilibrium	x	$0.1 - 2x$ mol dm^{-3}

$$\frac{(0.1 - 2x)^2}{x} = 2.19 \times 10^{-3}$$

$$4x^2 - 0.402\ 19x + 0.01 = 0$$

$$x = \frac{0.402\ 19 \pm \sqrt{0.001\ 76}}{8} = \frac{0.402\ 19 \pm 0.041\ 95}{8}$$

$= 0.0555$ or 0.04503 mol

Only the second is possible, and this leads to

Dimer: 0.045 mol dm^{-3}; Monomer: 0.01 mol dm^{-3}

13. K_p at 3000 K $= \dfrac{(0.4)^2 \times 0.2}{(0.6)^2} = 0.0889$ atm

$\Delta G^\circ/\text{J mol}^{-1} = -8.314 \times 3000 \ln 0.0889$

$\Delta G^\circ = 60\ 370 \text{ J mol}^{-1} = 60.4 \text{ kJ mol}^{-1}$

14. (a) $\ln K_c = -\dfrac{2930}{8.314 \times 310.15} = -1.136$; $K_c = 0.321$

(b) $\ln K_c = +\dfrac{15\ 500}{8.314 \times 310.15} = 6.01$; $K_c = 408$

(c) $K_c = 0.321 \times 408 = 130.9$

$\Delta G^\circ = 2.93 - 15.5 = -12.6 \text{ kJ mol}^{-1}$

15. (a) $\Delta G^\circ = -2 \times 26.57$ kJ mol^{-1} = 53.14 kJ mol^{-1}

$K_P = \exp(-\Delta G^\circ/RT) = 2.04 \times 10^9$ atm^{-2}

(b) $\Delta G^\circ = -32.89 - 200.92 = -233.81$ kJ mol^{-1}

$K_P = 9.2 \times 10^{40}$ atm^{-2}

(c) $\Delta G^\circ = -32.89 - 68.12 = -101.01$ kJ mol^{-1}

$K_P = 5.0 \times 10^{17}$ atm^{-1}

(d) $\Delta G^\circ = -32.89 + (2 \times 50.75)$

= 68.61 kJ mol^{-1}

$K_P = 9.5 \times 10^{-13}$

16. (a) $\Delta G^\circ = -RT \ln K$

$= -8.314 \times 298.15 \ln 10^{-5}$

= 28 538 J mol^{-1} = 28.5 kJ mol^{-1}

$\Delta H^\circ = \Delta G^\circ + T\Delta S^\circ$ = 28 538 − 298.15 x 41.8

= 16 075 J mol^{-1} = 16.1 kJ mol^{-1}

(b)	CO	+	$H_2O \rightleftharpoons$	CO_2	+ H_2
Initially:	2		2	0	0 mol
At Equilibrium:	$2 - x$		$2 - x$	x	x mol

$$\frac{x^2}{(2 - x)^2} = 1.00 \times 10^{-5}$$

$$\frac{x}{2 - x} = 3.16 \times 10^{-3}; \; x = 6.30 \times 10^{-3}$$

The amounts are therefore

1.994 mol (CO); 1.994 mol (H_2O)

6.30×10^{-3} mol (CO_2); 6.30×10^{-3} mol (H_2)

17. For reaction (1)

$K_1 = \exp(-15\,700/8.310 \times 310.15)$

$= 2.27 \times 10^{-3}\ dm^3\ mol^{-1}$

For reaction (2)

$K_2 = \exp(31\,000/8.31 \times 310.15) = 1.66 \times 10^5\ mol\ dm^{-3}$

For the coupled reaction (3)

$K_3 = K_1K_2 = 377$

18. (a) Zero

(b) $\Delta G^\circ = 0;\ \Delta H^\circ > 0;\quad \therefore \Delta S^\circ > 0.$

(c) $K_c = K_p(RT)^{-\Sigma\nu} = (0.082\,06 \times 298.15)^{-1}$

$= 0.0409\ mol\ dm^{-3}$

$\Delta G^\circ = -8.314 \times 298.15 \ln 0.0409 = 7924\ J\ mol^{-1}$

$= 7.92\ kJ\ mol^{-1}$

(d) $K_p > 1$ atm

(e) $\Delta G^\circ < 0$

19. (a) Yes

(b) Yes

(c) $\Delta S^\circ > 0$

20. (a) $\Delta G^\circ = -32.98 - 68.12 = -101.01\ kJ\ mol^{-1}$

$\Delta H^\circ = -84.68 - 52.26 = -136.94\ kJ\ mol^{-1}$

$$\Delta S^\circ = \frac{-136\,940 + 101\,010}{298.15} = -120.5\ J\ K^{-1}\ mol^{-1}$$

Standard state is 1 atm

(b) $$\ln (K_p/atm^{-1}) = -\frac{\Delta G^\circ}{RT} = \frac{101\,010}{8.314 \times 298.15} = 40.75$$

$K_p = 4.98 \times 10^{17}\ atm^{-1}$

(c) $K_c = 4.98 \times 10^{17} \times 0.082\,06 \times 298.15$

$= 1.22 \times 10^{19}\ dm^3\ mol^{-1}$

(d) $\Delta G^\circ = -8.314 \times 298.15 \ln (1.22 \times 10^{19})$

$= 108\ 940 \text{ J mol}^{-1} = 108.94 \text{ kJ mol}^{-1}$

(e) $\Delta S^\circ = \dfrac{-136\ 940 + 108\ 940}{298.15} = -93.9 \text{ J K}^{-1} \text{ mol}^{-1}$

$\Delta G^\circ\ (100\,^\circ\text{C}) = -136\ 940 + (120.5 \times 373.15)$

$= -91\ 975 \text{ J mol}^{-1}$

$\ln (K_p/\text{atm}^{-1}) = \dfrac{91\ 975}{8.314 \times 373.15} = 29.65$

$K_p\ (100\,^\circ\text{C}) = 7.51 \times 10^{12} \text{ atm}^{-1}$

21. (a) $\Delta G^\circ = 2(-228.57) = -457.14 \text{ kJ mol}^{-1}$

$\Delta H^\circ = 2(-241.84) = -483.68 \text{ kJ mol}^{-1}$

$\Delta S^\circ = \dfrac{-483\ 680 + 457\ 140}{298.15} = -89.02 \text{ J K}^{-1} \text{ mol}^{-1}$

(b) $\ln (K_p/\text{atm}^{-1}) = \dfrac{457\ 140}{8.314 \times 298.15} = 184.4$

$K_p = 1.24 \times 10^{80} \text{ atm}^{-1}$

(c) $\Delta G^\circ\ (2000\,^\circ\text{C})/\text{J mol}^{-1}$

$= -483\ 680 + (89.02 \times 2273.15)$

$\Delta G^\circ = -281\ 320 \text{ J mol}^{-1} = -281.3 \text{ kJ mol}^{-1}$

$\ln (K_p/\text{atm}^{-1}) = \dfrac{281\ 320}{8.314 \times 2273.15} = 14.89$

$K_p = 2.92 \times 10^{6} \text{ atm}^{-1}$

22. (a) $\Delta S^\circ = \dfrac{\Delta H^\circ - \Delta G^\circ}{T} = \dfrac{-20\ 100 + 31\ 000}{310.15}$

$= 35.1 \text{ J K}^{-1} \text{ mol}^{-1}$

(b) $\ln K = \dfrac{31\ 000}{8.314 \times 310.15} = 12.02$

$K_c = 1.66 \times 10^{5} \text{ mol dm}^{-3}$

(c) $\Delta G^\circ\ (25\,^\circ\text{C}) = -20\ 100 - (298.15 \times 35.1)$

$$= -30\ 570 \text{ J mol}^{-1}$$

$$\ln K_c = \frac{30\ 570}{8.314 \times 298.15} = 12.33$$

$$K_c = 2.27 \times 10^5 \text{ mol dm}^{-3}$$

23. (a) $\Delta H° = -165.98 + 146.44 = -19.54 \text{ kJ mol}^{-1}$

$\Delta S° = 306.4 - 349.0 = -42.6 \text{ J K}^{-1} \text{ mol}^{-1}$

$\Delta G° = -19\ 540 + (42.6 \times 298.15)$

$= -6836 \text{ J mol}^{-1} = -6.84 \text{ kJ mol}^{-1}$

(b) $\ln K = \dfrac{6839}{8.314 \times 298.15} = 2.759; \quad K = 15.78$

If partial pressure of neopentane = x atm

partial pressure of n-pentane = $(1 - x)$ atm

$$\frac{x}{1 - x} = 15.78$$

$x = 15.78 - 15.78\ ; \quad x = 0.940$

$1 - x = 0.060$

Thus P(neopentane) = 0.940 atm and

P(n-pentane) = 0.060 atm

24. (a) Slope of plot of $\ln K_c$ against $1/T$ is

$$\left(\frac{\ln 3}{\frac{1}{298.15} - \frac{1}{313.15}}\right) = -\left(\frac{1.0986}{1.607 \times 10^{-4}}\right) = -6836$$

$\Delta H° = 6836 \times 8.314 = 56\ 840 \text{ J mol}^{-1}$

$= 56.8 \text{ kJ mol}^{-1}$

(b) $-56.8 \text{ kJ mol}^{-1}$

25. Slope of plot of $\ln K_c$ against $1/T$ is

$$-\frac{\ln 1.45}{\dfrac{1}{298.15} - \dfrac{1}{303.15}} = -\frac{0.372}{5.532 \times 10^{-5}} = -6724.6 \text{ K}$$

$$\Delta H° = 6724.6 \times 8.314 = 55\ 910 \text{ J mol}^{-1}$$

$$= 55.9 \text{ kJ mol}^{-1}$$

26. (a) At 400°C,

$$\log_{10}(K_P/\text{atm}) = 7.55 - \frac{4844}{673.15} = 0.354$$

$$K_P = 2.259 \text{ atm}$$

$$\Delta G°/\text{kJ mol}^{-1} = -8.314 \times 673.15 \ln 2.254$$

$$\Delta G° = -4548 \text{ J mol}^{-1} = -4.55 \text{ kJ mol}^{-1}$$

$$\Delta H° = 4844 \times 8.314 \times 2.303 \text{ J mol}^{-1}$$

$$= 92\ 750 \text{ J mol}^{-1}$$

$$= 92.75 \text{ kJ mol}^{-1}$$

$$\Delta S° = \frac{92\ 800 + 4548}{673.15} \text{ J K}^{-1} \text{ mol}^{-1}$$

$$= 144.6 \text{ J K}^{-1} \text{ mol}^{-1}$$

(b) $K_c = 2.259 \times (0.082\ 06 \times 673.15)^{-1}$

$$= 0.0409 \text{ mol dm}^{-3}$$

$$\Delta G°/\text{kJ mol}^{-1} = -8.314 \times 673.15 \ln 0.0409$$

$$\Delta G° = 17\ 890 \text{ J mol}^{-1} = 17.89 \text{ kJ mol}^{-1}$$

(c) I_2 + cyclopentene $\rightleftharpoons$ 2HI + cyclopentadiene

$0.1 - x$ $\quad$ $0.1 - x$ $\quad$ $2x$ $\quad$ x

$$\frac{4x^3}{(0.1 - x)^2} = 0.0409$$

For a very approximate solution, neglect x in comparison with 0.1:

$$4x^3 = 0.0409 \times (0.1)^2 = 4.09 \times 10^{-4}$$

$$x^3 = 1.0225 \times 10^{-4}$$

$$x = 0.0468$$

For a better solution, calculate $4x^3/(0.1 - x)^2$ at various x values:

x	0.0468	0.04	0.03	0.0350	0.0355
$\dfrac{4x^3}{(0.1 - x)^2}$	0.1449	0.0711	0.022	0.0406	0.0430

Additionally, ($x = 0.0351$)/value of term = 0.0411

$x = 0.0351$

Final concentrations are

$[I_2] = 0.0649$ M; [HI] = 0.0702 M

[cyclopentene] = 0.0649 M;

[cyclopentadiene] = 0.0351 M

27. $\Delta H^\circ = -283.66 + 110.54 = -173.12$ kJ mol^{-1}

$\Delta G^\circ = -166.36 + 137.15 = -29.21$ kJ mol^{-1}

$$\Delta S^\circ = \frac{-173\ 120 + 29\ 210}{298.15} = -482.7 \text{ J K}^{-1} \text{ mol}^{-1}$$

$$\ln K_P = \frac{29\ 210}{8.314 \times 298.15} = 11.78$$

$$K_P = 1.31 \times 10^5 \text{ atm}^{-3}$$

28. $\Delta H^\circ = -207.4 + 104.6 = -102.8$ kJ mol^{-1}

$\Delta G^\circ = -111.3 + 37.2 = -74.1$ kJ mol^{-1}

$$\Delta S^\circ = \frac{-102\ 800 + 74\ 100}{298.15} = -98.3 \text{ J K}^{-1} \text{ mol}^{-1}$$

29. (a) $K_c = 95/5 = 19$

$\Delta G^\circ/\text{J mol}^{-1} = -8.314 \times 298.15 \ln 19$

$\Delta G^\circ = -7298 \text{ J mol}^{-1} = -7.30 \text{ kJ mol}^{-1}$

(b) $\Delta G/\text{J mol}^{-1} = -7298 + 8.314 \times 298.15 \ln \frac{10^{-4}}{10^{-2}}$

$\Delta G = (-7298 - 11\ 415) \text{ J mol}^{-1} = -18\ 710 \text{ J mol}^{-1}$

$= -18.7 \text{ kJ mol}^{-1}$

The reaction will therefore go from left to right.

30. (a) $\Delta H^\circ = -110.54 - 241.84 + 393.51 = 41.13 \text{ kJ mol}^{-1}$

$\Delta G^\circ = -137.15 - 228.57 + 394.34 = 28.62 \text{ kJ mol}^{-1}$

$\Delta S^\circ = \frac{41\ 130 - 28\ 620}{298.15} = 41.96 \text{ J K}^{-1} \text{ mol}^{-1}$

(b) $\ln K_P = -\frac{28\ 620}{8.314 \times 298.15} = -11.55$

$K_P = 9.64 \times 10^{-6}$

(c) From the data in Table 2.1,

$\Delta d = 28.41 + 30.54 - 44.22 - 27.28$

$= -12.55 \text{ J K}^{-1} \text{ mol}^{-1}$

$\Delta e = 10^{-3} (4.10 + 10.29 - 8.79 - 3.26)$

$= 2.34 \times 10^{-3} \text{ J K}^{-2} \text{ mol}^{-1}$

$\Delta f = 10^{4} (-4.6 + 0 + 86.2 - 5.0)$

$= 76.6 \times 10^{4} \text{ J K mol}^{-1}$

Then, from Eq. 2.52,

$$\Delta H^\circ_T/\text{J mol}^{-1} = 41\ 130 - 12.55[(T/\text{K}) - 298.15] + 1.17 \times 10^{-3}[(T/\text{K})^2 - 298.15^2] + 76.6 \times 10^{4}[\frac{1}{T/\text{K}} - \frac{1}{298.15}]$$

$= 42\ 199 - 12.55(T/\text{K}) + 1.17 \times 10^{-3}(T/\text{K})^2 +$

$$76.6 \times 10^4/(T/\mathrm{K})$$

(d) $$\ln K = \int \frac{\Delta H^\circ}{RT^2}\, dT$$

$$= \frac{1}{8.314} \int \left(\frac{42\ 199}{(T/\mathrm{K})^2} - \frac{12.55}{T/\mathrm{K}} + 1.17 \times 10^{-3} + \frac{76.6 \times 10^4}{(T/\mathrm{K})^3} \right) d(T/\mathrm{K})$$

$$= \frac{1}{8.314} \left(- \frac{42\ 199}{(T/\mathrm{K})} - 12.55 \ln (T/\mathrm{K}) + 1.17 \times 10^{-3} (T/\mathrm{K}) - \frac{38.3}{(T/\mathrm{K})^2} \right) + \mathrm{I}$$

$$= - \frac{5076}{T/\mathrm{K}} - 1.51 \ln (T/\mathrm{K}) + 1.41 \times 10^{-4} (T/\mathrm{K}) - \frac{4.61}{(T/\mathrm{K})^2} + \mathrm{I}$$

I is obtained from the fact that at 298.15,

$\ln K = -11.55$

$$\mathrm{I} = -11.55 + 17.02 + 8.60 - 0.042 + 5.19 \times 10^{-5}$$

$$= 14.03$$

$$\ln K = 14.03 - \frac{5076}{T/\mathrm{K}} - 1.51 \ln (T/\mathrm{K}) + 1.41 \times 10^{-4} (T/\mathrm{K}) - \frac{4.61}{(T/\mathrm{K})^2}$$

(e) $\ln K_P$ at 1000 K

$$= 14.03 - \frac{5076}{1000} - 1.51 \ln 1000 + 0.141 - \frac{4.61}{1000^2}$$

$$= -1.62$$

$$K_P\ (1000\ \mathrm{K}) = 0.20$$

31. Partial pressures at 1395 K are: CO: 0.000 140 atm; CO_2: (1 - 0.000 140) atm; O_2: 0.000 070 atm

$$K_P = \frac{(0.000\ 140)^2 \times 0.000\ 070}{0.9986} = 1.372 \times 10^{-12}\ \mathrm{atm}$$

At 1443 K,

$$K_P = \frac{(0.000\ 250)^2 \times 0.000\ 0125}{0.999\ 75} = 7.814 \times 10^{-12}\ \text{atm}$$

At 1498 K,

$$K_P = \frac{(0.000\ 471)^2 \times 0.000\ 2355}{0.999\ 539} = 5.227 \times 10^{-11}\ \text{atm}$$

Then

T/K	$10^{12} K_P$/atm	$10^4/(T/\text{K})$	$\ln(10^{12} K_P/\text{atm})$
1395	1.372	7.168	0.3163
1443	7.814	6.930	2.0559
1498	52.27	6.676	3.956

From a plot of $\ln(K_P/\text{atm})$ against $1/(T/\text{K})$, slope $= -7.33 \times 10^4$ K and $\Delta H° = -R \times \text{slope} = 609\ \text{kJ mol}^{-1}$.

$$\Delta G°(1395\ \text{K}) = -8.314 \times 1395 \ln (1.372 \times 10^{-12})$$

$$= 316.8\ \text{kJ mol}^{-1}$$

$$\Delta S° = \frac{\Delta H° - \Delta G°}{T} = 210\ \text{J K}^{-1}\ \text{mol}^{-1}$$

(standard state 1 atm).

32. Suppose that there are present x mol of I_2 and y mol of I:

$$x + \frac{y}{2} = 1.958 \times 10^{-3}$$

$$P = \frac{(x + y)\ \text{mol}\ RT}{V}$$

At 800°C

$$\frac{558.0}{760.0}\ \text{atm} = \frac{(x + y)\ 0.082\ 05\ \text{dm}^3\ \text{atm K}^{-1}\ \text{mol}^{-1}}{249.8 \times 10^{-3}\ \text{dm}^3} \times$$

$$1073.15\ \text{K}$$

$$x + y = 2.0829 \times 10^{-3}$$

$$\frac{y}{2} = 0.1249 \times 10^{-3};\ y = 2.498 \times 10^{-4}$$

$$x = 1.833 \times 10^{-3}$$

$\alpha = 0.0638$

At 1000°C, $x + y = 2.3535 \times 10^{-3}$

$\frac{y}{2} = 3.955 \times 10^{-4} \qquad y = 7.91 \times 10^{-4}$

$x = 1.5625 \times 10^{-3}; \ \alpha = 0.202$

At 1200°C, $x + y = 2.7715 \times 10^{-3}$

$\frac{y}{2} = 8.135 \times 10^{-4}; \ y = 1.627 \times 10^{-3}$

$x = 1.1445 \times 10^{-3}; \ \alpha = 0.415$

(a) α = 0.0638, 0.202, 0.415 at the 3 temperatures

(b) At 800°C,

$$K_c = \frac{(2.498 \times 10^{-4}\ \text{mol})^2}{1.833 \times 10^{-3}\ \text{mol} \times 0.2498\ \text{dm}^3}$$

$$= 1.363 \times 10^{-4}\ \text{mol dm}^{-3}$$

At 1000°C

$$K_c = \frac{(7.91 \times 10^{-4})^2}{1.5625 \times 10^{-3} \times 0.2498}$$

$$= 16.03 \times 10^{-4}\ \text{mol dm}^{-3}$$

At 1200°C

$$K_c = \frac{(1.627 \times 10^{-3})^2}{1.1445 \times 10^{-3} \times 0.2498}$$

$$= 92.59 \times 10^{-4}\ \text{mol dm}^{-3}$$

(c) $K_P = K_c\,(RT)^{\Sigma\nu}$ (Eq. 4.26)

$= K_c RT$ since $\Sigma\nu = 1$

At 800°C, $K_P = 1.363 \times 10^{-4}\ \text{mol dm}^{-3} \times$

$1000\ \text{dm}^3\ \text{m}^{-3} \times$

$8.314\ \text{J K}^{-1}\ \text{mol}^{-1} \times 1073.15\ \text{K}$

= 1.216 kPa = 9.12 mmHg = 0.0120 atm

At 1000°C, K_P = 16.97 kPa

At 1200°C, K_p = 98.00 kPa

(d)

T/K	$10^4 K_c$/mol dm^{-3}	$10^4/(T/\mathrm{K})$	$\ln \frac{10^4 K_c}{\mathrm{mol\ dm^{-3}}}$
1073.15	1.363	9.318	0.310
1273.15	16.03	7.855	2.774
1473.15	92.59	6.788	4.528

Slope of a plot of $\ln(K_c/\mathrm{mol\ dm^{-3}})$ against $1/(T/\mathrm{K})$ is $-1.66 \times 10^4\ \mathrm{K^{-1}}$;

$\Delta U° = 138\ \mathrm{kJ\ mol^{-1}}$

$\Delta H° = \Delta U° + RT = 138\ 000 + (8.314 \times 1273.15)$

$= 148\ 600\ \mathrm{J\ mol^{-1}} = 149\ \mathrm{kJ\ mol^{-1}}$

(e) At 1000°C, K_p = 16.97 kPa = 0.167 atm.

$\Delta G° = -(8.314 \times 1273.15\ \mathrm{J\ mol^{-1}})\ \ln(0.167/\mathrm{atm})$

$= 18\ 914\ \mathrm{J\ mol^{-1}}$

$= 18.9\ \mathrm{kJ\ mol^{-1}}$

(standard state: 1 atm)

$$\Delta S° = \frac{\Delta H° - \Delta G°}{T} = 101.9\ \mathrm{J\ K^{-1}\ mol^{-1}}$$

33. If the concentration of M is [M], that of sites occupied and unoccupied, is n[M]. The association reactions may be formulated in terms of the sites, S:

$$\mathrm{S + A \underset{}{\overset{K_s}{\rightleftharpoons}} SA}$$

$$K_s = \frac{[\mathrm{SA}]}{[\mathrm{S}][\mathrm{A}]}$$

where [S] is the concentration of unoccupied sites and [SA] the concentration of occupied sites. The total concentration of sites, n[M], is

$$n[\mathrm{M}] = [\mathrm{S}] + [\mathrm{SA}] = [\mathrm{SA}]\left\{\frac{1}{K_s[\mathrm{A}]} + 1\right\}$$

The average number of occupied sites per molecule is the total concentration of occupied sites divided by the total concentration of M:

$$\bar{\nu} = \frac{[\mathrm{SA}]}{[\mathrm{M}]} = \frac{n}{\dfrac{1}{K_s[\mathrm{A}]} + 1} = \frac{n\ K_s[\mathrm{A}]}{1 + K_s[\mathrm{A}]}$$

34. The total concentration of the molecule M is

$$[\mathrm{M}]_o = [\mathrm{M}] + [\mathrm{MA}] + [\mathrm{MA}_2] + \ldots.\ [\mathrm{MA}_n]$$

The total concentration of occupied sites is the total concentration of bound A molecules:

$$[\mathrm{A}]_b = [\mathrm{MA}] + 2[\mathrm{MA}_2] + \ldots + n[\mathrm{MA}_n]$$

Expressing every term in terms of [A]:

$$[\mathrm{M}]_o = [\mathrm{M}]\left\{1 + K_1[\mathrm{A}] + K_1K_2[\mathrm{A}]^2 + \ldots + (K_1K_2 \ldots.\ K_n)[\mathrm{A}]^n\right\}$$

$$[\mathrm{A}]_b = [\mathrm{M}]\left\{K_1[\mathrm{A}] + 2K_1K_2[\mathrm{A}]^2 + \ldots\ n(K_1K_2 \ldots.\ K_n)[\mathrm{A}]^n\right\}$$

Thus

$$\bar{\nu} = \frac{[\mathrm{A}]_b}{[\mathrm{M}]_o} = \frac{K_1[\mathrm{A}] + 2K_1K_2[\mathrm{A}]^2 + \ldots + n(K_1K_2 \ldots K_n)[\mathrm{A}]^n}{1 + K_1[\mathrm{A}] + K_1K_2[\mathrm{A}]^2 + \ldots + (K_1K_2 \ldots K_n)[\mathrm{A}]^n}$$

35. With $K_1 = n\ K_s$, $K_2 = (n - 1)K_s/2$, $K_3 = (n - 2)K_s/3$, ..., $K_n = K_s/n$, the preceding equation becomes

$$\bar{\nu} = \frac{n\ K_s[\mathrm{A}] + n(n - 1)K_s^2[\mathrm{A}]^2 + \ldots + n\ K_s^n[\mathrm{A}]^n}{1 + n\ K_s[\mathrm{A}] + n(n - 1)K_s^2[\mathrm{A}]^2/2 + \ldots + K_s^n[\mathrm{A}]^n}$$

$$= \frac{n\ K_s[\mathrm{A}]\left\{1 + (n - 1)K_s[\mathrm{A}] + \ldots + K_s^{\,n-1}[\mathrm{A}]^{n-1}\right\}}{1 + n\ K_s[\mathrm{A}] + n(n - 1)K_s^{\,2}[\mathrm{A}]^2/2 + \ldots + K_s^{\,n}[\mathrm{A}]^n}$$

The coefficients are the binomial coefficents:

$$\bar{\nu} = \frac{n\ K_s[\mathrm{A}](1 + K_s[\mathrm{A}])^{n-1}}{(1 + K_s[\mathrm{A}])^n}$$

$$= \frac{n\ K_s[\mathrm{A}]}{1 + K_s[\mathrm{A}]}$$

which is the expression obtained in Problem 33.

To test the equation, plot $1/\bar{\nu}$ against $1/[\mathrm{A}]$:

$$\frac{1}{\bar{\nu}} = \frac{1}{n} + \frac{1}{n\ K_s[\mathrm{A}]}$$

One of the intercepts is $1/n$. Alternatively, plot $\bar{\nu}$ against $\bar{\nu}/[\mathrm{A}]$:

$$\bar{\nu} = n - \frac{\bar{\nu}}{K_s[\mathrm{A}]}$$

36. If K_n is very much larger than K_1, K_2, etc., the equation obtained in Problem 34 reduces as follows:

$$\bar{\nu} = \frac{n(K_1K_2\ \ldots\ldots\ K_n)[\mathrm{A}]^n}{1 + (K_1K_2\ \ldots\ldots\ K_n)[\mathrm{A}]^n}$$

$$= \frac{n\ K[\mathrm{A}]^n}{1 + K[\mathrm{A}]^n}$$

where $K = K_1K_2\ \ldots\ldots\ K_n$ is the overall equilibrium constant for the binding of n molecules:

$$n\ \mathrm{A} + \mathrm{M} \overset{K}{\rightleftharpoons} \mathrm{MA}_n$$

The fraction if sites occupied, θ, is $\bar{\nu}/n$:

$$\theta = \frac{K[\mathrm{A}]^n}{1 + K[\mathrm{A}]^n}$$

or

$$\frac{\theta}{1 - \theta} = K[\mathrm{A}]^n$$

The slope of a plot of ln $\{\theta/(1 - \theta)\}$ against [A] is thus n. If the sites are identical and independent (Problem 33) the slope is 1. Intermediate behavior can give nonlinear plots; the maximum slope of a Hill plot cannot be greater than n.

37. Values of ΔG° and of $K[= \exp(-\Delta G^\circ/RT)^u]$ are

Temperature $\theta/°C$	T/K	ΔG°/kJ mol^{-1}	K
40.0	313.15	3.98	0.22
42.0	315.15	2.20	0.43
44.0	317.15	0.42	0.85
46.0	319.15	-1.36	1.67
48.0	321.15	-3.14	3.24
50.0	223.15	-4.93	6.27

$\Delta G^\circ = \Delta H^\circ - T\Delta S^\circ = 0$ when T = 317.6 K = 44.4°C

CHAPTER 5: WORKED SOLUTIONS
PHASES AND SOLUTIONS

1. $\ln \dfrac{P_2}{P_1} = \dfrac{-\Delta_{vap}H_m}{R}\left(\dfrac{1}{T_2} - \dfrac{1}{T_2}\right) = \dfrac{\Delta_{vap}H_m}{R}\left(\dfrac{T_2 - T_1}{T_2T_1}\right)$

$$\ln \frac{101.3}{3.17} = \frac{\Delta_{vap}H_m/\text{J mol}^{-1}}{8.314}\left(\frac{373.15 - 298.15}{373.15 \cdot 298.15}\right)$$

$$\Delta_{vap}H_m/\text{J mol}^{-1} = 3.46\ (8.314)\left(\frac{111\ 254}{75}\right)$$

$$= 42.67\ \text{kJ mol}^{-1}$$

$$\Delta_{vap}H_m = 42.7\ \text{kJ mol}^{-1}$$

The CRC Handbook gives 40.57 kJ mol^{-1}

2. The molar volume of iodine is

$$V_m = \frac{0.253\ 81\ \text{kg mol}^{-1}}{4.93\ \text{kg dm}^{-3}} = 51.48 \times 10^{-3}\ \text{dm}^3\ \text{mol}^{-1}$$

$$= 51.48 \times 10^{-6}\ \text{m}^3\ \text{mol}^{-1}$$

Then from Eq. 5.23,

$$\ln \frac{P_1^g}{P_2^g} = \frac{V_m(P_1 - P_2)}{RT}$$

$$\ln \frac{P_1^g}{P_2^g} = \frac{51.48 \times 10^{-6}(101.3 \times 10^6 - 101\ 300)}{8.314 \times 313.15} = 2.001$$

$$P_1^g/P_2^g = 7.40$$

At 101.3 kPa the pressure is 133 Pa. Therefore, at 101.3 x 10^3 kPa the pressure is 7.40 x 133 Pa = 984 Pa

3. The cubic expansion coefficient is a second order transition since it can be expressed as

$$\alpha = \frac{1}{V}\left[\frac{-\partial}{\partial T}\left(\frac{\partial G}{\partial P}\right)_T\right]_P$$

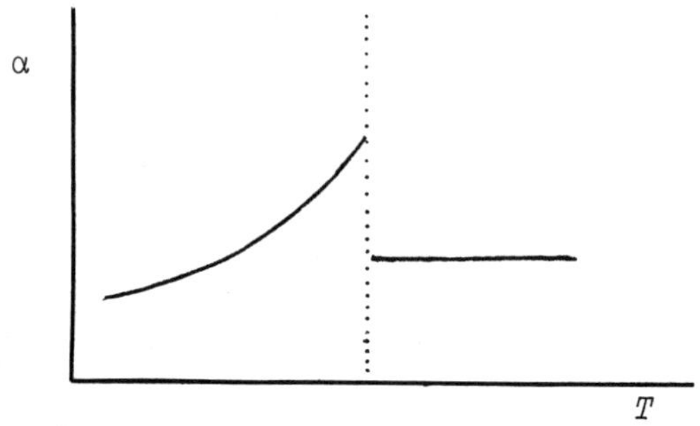

4. From Eq. 5.16,

$$\Delta_{vap}H_m = \left(\frac{1}{T_1} - \frac{1}{T_2}\right)^{-1} R \ln \frac{P_2}{P_1}$$

$$\Delta_{vap}H_m/\text{J mol}^{-1} = (4.97 \times 10^{-4})^{-1}\ 8.314 \ln \frac{31.860}{1.940}$$

$$= (2012)(8.314)(2.799) = 46\ 822$$

$$\Delta_{vap}H_m = 46\ 822 \text{ J mol}^{-1} = 46.8 \text{ kJ mol}^{-1}$$

5. $$\ln \frac{P_2}{P_1} = \Delta_{vap}H_m\left(\frac{1}{T_2} - \frac{1}{T_1}\right) = \frac{\Delta_{vap}H_m}{R}\frac{(T_2 - T_1)}{T_2T_1}$$

At $T_1 = 286 + 273.15 = 559.15$ K $P_1 = 101\ 325$ Pa

At $T_2 = 145 + 273.15 = 418.15$ $P_2 = 1866.5$ Pa

$$\ln \frac{1866.5}{101\ 325} = \frac{\Delta_{vap}H_m/\text{J mol}^{-1}}{8.314}\left[\frac{418.15 - 559.15}{(418.15)(559.15)}\right]$$

$$\Delta_{vap}H_m/\text{J mol}^{-1} = +3.994\ \frac{(8.314)(233\ 808)}{141}$$

$$\Delta_{vap}H_m = 55.06 \text{ kJ mol}^{-1}$$

The CRC Handbook value is 71.02 kJ mol^{-1}. The error is large but considering the relative molecular mass of the compound, its high boiling point, and the wide range of T and P involved in the calculation, it is not surprising that the error is so large.

6. Using Trouton's Rule

$$\Delta_{vap}H_m = 88\ \text{J K}^{-1}\ \text{mol}^{-1}\ 342.10\ \text{K} = 30.105\ \text{kJ mol}^{-1}$$

$$\%\ \text{error} = \frac{31.912 - 30.105}{31.912} \times 100 = 5.7\%$$

7. 2-Propanone is not particularly associated. Therefore, using $\Delta T \approx 9.3 \times 10^{-4}\ T_b \Delta P$, we have

$$\Delta T \approx -9.3 \times 10^{-4} (329.35)(2.825) = -0.865\ \text{K}$$

Thus T_b at 98.5 kPa is 329.35 - 0.86 = 328.49 K

8. Since water is associated, the numerical value 7.5×10^{-4} should be used in Eq. 5.20. T_b = 373.15 and ΔP = 102.7 - 101.3 = 1.4 kPa, and thus

$\Delta T = 7.5 \times 10^{-4} \times 373.15 \times 1.4 = 0.39$ K

The calculated boiling point at 101.3 is thus 373.52 - 0.39 = 373.11 K, an error of only 0.04 K.

9. We start with the Clapeyron equation, Eq. 5.9 and substitute $\frac{RT + M}{P}$ for V; this gives, after rearrangement

$$\frac{dP}{P} = \frac{\Delta_{vap}H_m}{T(RT + M)}$$

Since $\frac{1}{T(RM + M)} = \frac{1}{MT} - \frac{R}{M(RT + M)}$

$$\frac{dP}{P} = \frac{\Delta_{vap}H_m}{m}\frac{dT}{T} - \frac{R}{M}\frac{\Delta_{vap}H_m dT}{(RT + M)}$$

Integration with $\Delta_{vap}H_m$ constant gives

$$\ln \frac{P_2}{P_1} = \frac{\Delta_{vap}H_m}{M} \ln \frac{T_2}{T_1} - \frac{\Delta_{vap}H_m}{M} \ln \frac{RT_2 + M}{RT_1 + M}$$

$$\ln \frac{P_2}{P_1} = \frac{\Delta_{vap}H_m}{M} \ln \frac{T_2}{T_1}\left(\frac{RT_1 + M}{RT_2 + M}\right)$$

10. $P_T = 0.60(3.572) + 0.40(9.657)$

$= 2.143 + 3.863 = 6.006$ kPa

$$x^{vap}_{\text{toluene}} = \frac{2.143}{6.006} = 0.357$$

11. The molality m_2 is the amount of solute divided by the mass of solvent. If W_1 is the mass of solvent, the solution contains

$m_2 W_1$ mol of solute and W_1/M_1 mol of solvent

The mole fraction is thus

$$x_2 = \frac{m_2 W_1}{W_1/M_1 + m_2 W_1} = \frac{m_2 M_1}{1 + m_2 M_1}$$

If we divide each term by its SI unit,

$$x_2 = \frac{(m_2/\text{mol kg}^{-1})(M_1/\text{kg mol}^{-1})}{1 + (m_2/\text{mol kg}^{-1})(M_1/\text{kg mol}^{-1})}$$

The customary unit for molar mass M_1 is g mol^{-1}, and we then obtain

$$x_2 = \frac{(m_2/\text{mol kg}^{-1})(M_1/1000 \text{ g mol}^{-1})}{1 + (m_2/\text{mol kg}^{-1})(M_1/1000 \text{ g mol}^{-1})}$$

$$= \frac{(m_2/\text{mol kg}^{-1})(M_1/\text{g mol}^{-1})}{1000 + (m_2/\text{mol kg}^{-1})(M_1/\text{g mol}^{-1})}$$

If the solution is sufficiently dilute the expression approximates to

$$x_2 = (m_2/\text{mol kg}^{-1})(M_1/\text{g mol}^{-1})/1000$$

12. If V is the volume of solution, concentration of solute = c_2; amount of solute = Vc_2; mass of solute = Vc_2M_2 where M_2 is the molar mass of solute.

Mass of solution = $V\rho$; Mass of solvent = $V\rho - Vc_2M_2$

Amount of solvent = $(V\rho - Vc_2M_2)/M_1$

The mole fraction is therefore

$$x_2 = \frac{Vc_2}{(V\rho - Vc_2M_2)/M_1 + Vc_2} = \frac{c_2M_1}{\rho + c_2(M_1 - M_2)}$$

Dividing each term by its SI unit,

$$x_2 = \frac{(c_2/\text{mol m}^{-3})(M_1/\text{kg mol}^{-1})}{\rho/\text{kg m}^{-3} + (c_2/\text{mol m}^{-3})[(M_1 - M_2)/\text{kg mol}^{-1}]}$$

The more customary units are:

for concentration: mol dm^{-3}; for molar mass:

g mol^{-1}; for density: kg dm^{-3} $\equiv$ g cm^{-3}

Then

$$x_2 = \frac{(1000\ c_2/\text{mol dm}^{-3})(M_1/1000\ \text{g mol}^{-1})}{1000\ \rho/\text{g cm}^{-3} + (1000\ c_2/\text{mol dm}^{-3})\dfrac{(M_1 - M_2)}{1000\ \text{g mol}^{-1}}}$$

$$= \frac{(c_2/\text{mol dm}^{-3})(M_1/\text{g mol}^{-1})}{1000\ \rho/\text{g cm}^{-3} + (c_2/\text{mol dm}^{-3})\dfrac{(M_1 - M_2)}{\text{g mol}^{-1}}}$$

If the solution is sufficiently dilute the density of the solution is approximately that of the pure solvent, ρ_1, and the second term in the denominator can be neglected;

then

$$x_2 = c_2M_1/\rho_1 = \frac{(c_2/\text{mol dm}^{-3})(M_1/\text{g mol}^{-1})}{1000\ \rho_1/\ \text{g cm}^{-3}}$$

13. From the preceding problem,

amount of solute = Vc_2; mass of solvent = $V\rho - Vc_2M_2$

Thus the molality is

$$m_2 = \frac{c_2}{\rho - c_2 M_2}$$

and the concentration in terms of m_2 is

$$c_2 = \frac{m_2 \rho}{1 + m_2 M_2}$$

Dividing throughout by SI units,

$$c_2/\text{mol m}^{-3} = \frac{(m_2/\text{mol kg}^{-1})(\rho/\text{kg m}^{-3})}{1 + (m_2/\text{mol kg}^{-1})(M_2/\text{kg mol}^{-1})}$$

In terms of more usual units,

$$1000\ c_2/\text{mol dm}^{-3} = \frac{(m_2/\text{mol kg}^{-1})(1000\ \rho/\text{g cm}^{-3})}{1 + (m_2/\text{mol kg}^{-1})\left(M_2/1000\ \frac{\text{g}}{\text{mol}}\right)}$$

$$c_2/\text{mol dm}^{-3} = \frac{1000(m_2/\text{mol kg}^{-1})(\rho/\text{g cm}^{-3})}{1000 + (m_2/\text{mol kg}^{-1})(M_2/\text{g mol}^{-1})}$$

In dilute solution

$$c_2/\text{mol dm}^{-3} \approx (m_2/\text{mol kg}^{-1})(\rho_1/\text{g cm}^{-3})$$

where ρ_1 is the density of the solvent. For aqueous solutions $\rho_1 \approx 1\ \text{g cm}^{-3}$ and therefore the numerical values of the concentration and the molality, in the above units, are very similar.

14. From Eq. 5.27

$$V_{\text{NaCl}}/\text{cm}^3\ \text{mol}^{-1} = \left(\frac{\partial V}{\partial n_{\text{NaCl}}}\right)_{n_{H_2O}} = \frac{\partial V}{\partial m}$$

$$= 17.8213 + 1.74782\ m - 0.141675\ m^2$$

From Eq. 5.33

$$dV_{H_2O} = -\left(\frac{n_{\text{NaCl}}}{n_{H_2O}}\right) dV_{\text{NaCl}} = -\left(\frac{m}{55.508}\right) dV_{\text{NaCl}}$$

Integration from $m = 0$ to m gives

$$\int dV_{H_2O} = -\int\left(\frac{1.74782}{55.508}\ m - \frac{0.28335}{(55.508)}\ m^2\right)dm$$

$$V_{H_2O} - V^*_{H_2O} = \frac{1.74782}{2(55.508)}\ m^2 + \frac{0.28335}{3(55.508)}\ m^3$$

With $V^*_{H_2O} = 18.068\ \text{cm}^3$,

$$V_{H_2O}/\text{cm}^3\ \text{mol}^{-1} = 18.068 - 0.015\ 744\ m^2 + 0.001\ 7016\ m^3$$

15. We first develop an expression for $(\partial\rho/\partial n_2)n_1$. Since $x_2 = n_2/(n_1 + n_2)$,

$$\left(\frac{\partial x_2}{\partial n_2}\right)_{n_1} = \frac{n_1}{(n_1 + n_2)^2}$$

and $\left(\frac{\partial\rho}{\partial n_2}\right)_{n_1} = \frac{d\rho}{dx_2}\left(\frac{\partial x_2}{\partial n_2}\right)_{n_1} = \frac{n_1}{(n_1 + n_2)^2}\ \frac{d\rho}{dx_2}$

Substitution and division by $(n_1 + n_2)$ gives

$$V_2 = \frac{M_2}{\rho} - (M_1x_1 + M_2x_2)\ \frac{x_1}{\rho^2}\ \frac{d\rho}{dx_2}$$

16. Substitution of $x_2 = 0.100$ into the expression in Problem 15 gives $\rho = 0.970\ 609\ \text{kg dm}^{-3}$. Differentiating the ρ equation with respect to x_2 gives

$$\frac{d\rho}{dx_2} = -0.289\ 30 + 0.598\ 14\ x_2 - 1.826\ 28\ x_2^2 + 2.377\ 52\ x_2^3 - 1.029\ 05\ x_2^4$$

$$= -0.245\ 47\ \text{kg dm}^{-3}$$

With $M_1(H_2O) = 0.018\ 016\ \text{kg mol}^{-1}$ and $M_2(\text{methanol}) = 0.032\ 043\ \text{kg mol}^{-1}$, substitution with $x_1 = 0.900$ gives $V_2 = 0.037\ 57\ \text{dm}^3\ \text{mol}^{-1}$.

17. From Raoult's Law, we have for propylene dibromide (p),

$$P_p = x(\ell)P_p^* = 0.600(128)\ \text{mmHg}\ (133.33)\,\text{Pa/mmHg}$$

$$= 10.239\ \text{kPa}$$

For ethylene dibromide (e),

$$P_e = x(\ell)P_e^* = 0.40(172\ \text{mmHg})\ (133.32)\,\text{Pa/mmHg}$$

$$= 9.172\ \text{kPa}$$

The total pressure is $P_{total} = 10.239\ \text{kPa} + 9.172\ \text{kPa} = 19.411\ \text{kPa}$. The mole fraction in the vapor phase is given by $x = \dfrac{P}{P_{total}}$

For propylene dibromide

$$x_p(v) = \frac{10.239\ \text{kPa}}{19.411\ \text{kPa}} = 0.527$$

For ethylene dibromide

$$x_e(v) = \frac{9.172\ \text{kPa}}{19.411\ \text{kPa}} = 0.473$$

18. (a) Henry's Law applies to the individual gas components and we require the partial pressures of N_2 and O_2. Since the partial pressure is directly proportional to the mole fraction, for N_2 we have $P_{N_2} = 0.80$ atm; and for O_2 we have: $P_{O_2} = 0.20$ atm. From Eq. 5.25,

$$x(N_2) = \frac{P_{N_2}}{k'_{N_2}} = \frac{0.80\ \text{atm}}{7.58 \times 10^{+4}\ \text{atm}} = 1.06 \times 10^{-5}$$

$$x(O_2) = \frac{P_{O_2}}{k'_{O_2}} = \frac{0.20\ \text{atm}}{3.88 \times 10^{+4}\ \text{atm}} = 5.15 \times 10^{-6}$$

(b) Since the mole fractions are so small, if we use 1 mol of water as a reference, then there are (to a very good approximation) 1.06×10^{-5} mol of N_2 and 5.15×10^{-6} mol of O_2. The concentration calculation requires the volume of the solution, which is obtained from the density:

$$c(N_2) = \frac{1.06 \times 10^{-5} \text{ mol } (N_2)}{1 \text{ mol}(H_2O)[0.018 \text{ kg } H_2O/\text{mol}(H_2O)]} \times \frac{0.9982 \text{ kg } H_2O}{1 \text{ dm}^3} = 5.88 \times 10^{-4} \ M$$

$$c(O_2) = \frac{5.15 \times 10^{-6} \text{ mol } (O_2)}{1 \text{ mol}(H_2O)[0.018 \text{ kg } H_2O/\text{mol}(H_2O)]} \times \frac{0.9982 \text{ kg } H_2O}{1 \text{ dm}^3} = 2.86 \times 10^{-4} \ M$$

19. From the relative vapor pressure lowering we have

$$\frac{P_1^* - P_1}{P_1^*} = \frac{0.041}{2.332} = \frac{W_2/M_2}{(W_1/M_1) + (W_2/M_2)}$$

$$= \frac{18.14/M_2}{100.0/18.12 + 18.04/M_2}$$

$$0.0176 = \frac{18.04/M_2}{5.55 + 18.04/M_2}$$

$$M_2 = \frac{18.04 - 0.32}{0.0976} = 181.6$$

The correct value is 182.18.

20. The vapor pressure has been reduced from 40.00 to 26.66 kPa, so that the mole fraction of the solvent is given by Raoult's Law as

$$x_1 = \frac{P_1}{P_1^*} = \frac{26.66 \text{ kPa}}{40.00 \text{ kPa}} = \frac{2}{3}$$

Let the amount of solute be n_2 mol; since $n_1 = 1$ mol, the mole fraction of solvent is

$$x_1 = \frac{1}{1 + n_2} = \frac{2}{3}$$

Consequently, $n_2 = 1/2$. Since 0.080 kg is half a mole, the molar mass is 0.160 kg mol^{-1} = 160 g mol^{-1}

21. From Eq. 5.74, $P_1^* - P_1 = x_2 P_1^*$

$13.3 - 12.6 = x_2(13.3)$ $x_2 = 0.053$

From Eq. 5.80

$$x_2 = \frac{W_2/M_2}{W_1/M_1 + W_2/M_2} = \frac{1}{(M_2W_1/M_1W_2) + 1}$$

$$\frac{M_2W_1}{M_1W_2} = \frac{1}{x_2} - 1 = 18.0; \quad \frac{M_2}{M_1} = 18.0 \times \frac{1.00}{10.00} = 1.80$$

22. $\Delta_f T = K_f m_2 = 1.5$ K

$1.5 = 7.0\ m_2/\text{mol kg}^{-1}$; $m_2 = 0.2143$ mol kg^{-1}

$$\text{mol \% impurity} = \frac{0.2143 \times 100}{0.2143 + 1000/128} = \frac{21.43}{8.027}$$

$$= 2.67\%$$

100.00% - 2.67% = 97.33 mol % pure

23. $a_1 = \dfrac{6.677}{9.657} = 0.69$ $\qquad a_2 = \dfrac{1.214}{3.572} = 0.34$

$f_1 = \dfrac{0.69}{0.67} = 1.03$ $\qquad f_2 = \dfrac{0.34}{0.33} = 1.03$

24. Amount of NaCl = 11.5 g/58.5 g mol^{-1} = 0.197 mol

Amount of H_2O = 100 g/18 g mol^{-1} = 5.555 mol

$x_1 = \dfrac{5.555}{5.752} = 0.966,$ $\qquad x_2 = 0.034$

$a_1 = \frac{95.325}{101.325} = 0.941 \qquad f_1 = \frac{0.941}{0.966} = 0.974$

25. From Figure 5.13 the maximum value of the entropy is $5.76\ J\ K^{-1}\ mol^{-1}$. Then $\Delta_{mix}G^{id} = -300(5.76)$ $= -1728\ J\ mol^{-1}$. Therefore the Gibbs energy of mixing ranges from 0 to $-1.73\ kJ\ mol^{-1}$ for a 50-50 mixture. Since this is a rather small driving force, in a non-ideal solution where $\Delta_{mix}H = 0$, the value of $\Delta_{mix}H$ must be negative or only slightly positive for mixing to occur.

26. $a_{H_2O} = \frac{P_{H_2O}}{P^*_{H_2O}} = \frac{2.269\ kPa}{2.339\ kPa} = 0.9700$

Since $a_{H_2O} = f_{H_2O}x_{H_2O}$

$f_{H_2O} = \frac{a_{H_2O}}{x_{H_2O}} = \frac{0.9700}{0.990} = 0.980$

27. Rewriting Eq. 5.25 as $P_2(k'')^{-1} = c_2$, the values for N_2 and O_2 may be substituted directly. We make the assumption that N_2 gives rise to 80% of the pressure and O_2 is responsible for the other 20%. Then,

$c(N_2) = 0.80(101\ 325)\ Pa\ 2.17 \times 10^{-8}\ mol\ dm^{-3}\ Pa^{-1}$

$= 1.76 \times 10^{-4}\ mol\ dm^{-3}$

$c(O_2) = 0.20(101\ 325)\ Pa\ 1.02 \times 10^{-8}\ mol\ dm^{-3}\ Pa^{-1}$

$= 2.06 \times 10^{-4}\ mol\ dm^{-3}$

The total concentration is $1.966 \times 10^{-3}\ mol\ dm^{-3}$. This value of the concentration approaches the value of the molality. We may then use the molal freezing point depression expression. The result is

$\Delta_{fus}T = -(1.86)(0.001\ 966) = -0.003\ 66\ \text{K}$

28. Since the molar mass is 60 g mol^{-1}, c = 0.1 mol dm^{-3}

$\pi = cRT = \frac{n}{V}RT = 0.1(8.314)(300)\ \text{kPa} = 249.4\ \text{kPa}$

29. $\Delta_{fus}T = K_f m_2 = 2.17 \times 1.50 = 3.25\ \text{K}$

From Eq. 5.111, substituting a_1 for x_1 as discussed in the paragraph after Eq. 5.96, we have

$$\ln a = \frac{\Delta_{fus}H^*}{R}\left(\frac{1}{T_f^*} - \frac{1}{T}\right) = -\frac{\Delta_{fus}H^*\theta}{R\ T_f^*}$$

$$\ln a_1 = -\frac{6009.5(3.25)}{8.314(273.15)^2} = -0.0315;\ a = 0.9690$$

$$f_1 = \frac{a}{x} = \frac{0.969/55.6}{55.6 + 1.5} = \frac{0.9690}{0.9737} = 0.995.$$

30. The molality of the solute is calculated from

$\Delta_{vap}T = K_b m_2$

$$m = \frac{\Delta_{fus}T}{K_b} = \frac{0.9\ \text{K}}{6.26\ \text{K m}^{-1}} = 0.144\ \text{m}$$

$$= 0.144\ \text{mol kg}^{-1}$$

The mass of solute per kilogram of solvent is

$$\frac{0.000\ 85\ \text{kg solute}}{0.150\ \text{kg bromobenzene}} = 0.005\ 67$$

Then, for the solute

0.005 67 kg = 0.144 mol

and

$$M = \frac{0.005\ 67\ \text{kg}}{0.144\ \text{mol}} = 0.0394\ \text{kg mol}^{-1} = 39.4\ \text{g mol}^{-1}$$

31. For a dissociation $A_xB_y \rightleftharpoons xA^{z+} \quad yB^{z-}$

the initial molalities: $m \quad 0 \quad 0$ mol kg^{-1}

Molalities after dissociation: $m - \alpha m \quad x\alpha m \quad + y\alpha m$ mol kg^{-1}

Total molality = $m(1 - \alpha + x\alpha + y\alpha)$ mol kg^{-1}

Then $$i = \frac{\text{total molality}}{\text{initial molality}} = 1 - \alpha + x\alpha + y\alpha$$

If $\nu = x + y$, $i = 1 - \alpha + \alpha\nu = 1 - \alpha(1 - \nu)$

Then $$\alpha = \frac{i - 1}{\nu - 1}$$

From our problem $\Delta_{fus}T = 273.150 - 273.114 = 0.036$ K

$i = \Delta_{fus}T/K_f m = 0.036\ K/[1.86(\text{K}\ m^{-1})0.010(m)] = 1.94$

Since complete dissociation gives $\nu = 2$ for HCl,

$$\alpha = \frac{i - 1}{\nu - 1} = \frac{0.94}{1} = 0.94$$

The electrolyte therefore appears to be 94% dissociated. This extent of dissociation is only apparent because of the nonideality of the solution.

32. From Eq. 5.130, $\pi = \dfrac{n_2RT}{V}$,

Thus $$\pi = \frac{m_2/M_2}{V} RT$$

where m_2 is the mass of solute dissolved in volume V and M_2 is the molar mass. Since the equations are good only for dilute solutions we try to find a limiting value of M_2. From the above $M_2 = (RT/\pi)(m_2/V)$ and so $\lim \dfrac{m_2}{V} \to 0$ and $\lim \dfrac{m_2}{\pi V} \to 0$. A plot of $(m_2/\pi V)$/g dm^{-3} atm^{-1} against (m_2/V)/ g dm^{-3} gives a limiting value of $m_2/\pi V$. The values of

$(m_2/\pi V)/\text{g dm}^{-3}\ \text{atm}^{-1}$ corresponding to the listed values are

12.93	12.98	12.68	12.16	11.53	10.92

From the plot the limiting value of $m_2/\pi V$ is about 12.9 g dm^{-3} atm^{-1}. Thus

$$M_2 = RT\left(\frac{m_2}{V}\right)_0 = 0.0821\ \text{atm dm}^3\ \text{K}^{-1}\ \text{mol}^{-1}\ 293.15\ (\text{K}) \times 12.9\ (\text{g dm}^{-3}\ \text{atm}^{-1}) = 310\ \text{g mol}^{-1}$$

The molar mass of sucrose is 342 g mol^{-1}, so that there is an error of about 9%. Ignoring the lowest concentration point and extrapolating leads to an error of about 4%. However since the slope is expected to be fairly close to zero at infinite dilution a better result cannot be obtained without more low-concentration data.

33. What we have to do is to find the value of μ_A such that $\mu_A + \mu_B$ $(=G)$ is equal to the expression given in the problem.

(a) $$\mu_A = \left(\frac{\partial G}{\partial n_A}\right)_{n_B,T,P} = \mu_A^* + RT\ln x_A + RTn_A\left(\frac{\partial \ln x_A}{\partial n_A}\right)_{n_B} + RTn_B\left(\frac{\partial \ln x_B}{\partial n_A}\right)_{n_B} + \frac{c\ n_B(n_A + n_B)}{(n_A + n_B)^2} - \frac{c\ n_A n_B}{(n_A + n_B)^2}$$

Since $$\left(\frac{\partial \ln x_A}{\partial n_A}\right)_{n_B} = \left[\frac{\partial\left(\frac{n_A}{n_A + n_B}\right)}{\partial n_A}\right]_{n_B} = \frac{1}{n_A} - \frac{1}{(n_A + n_B)}$$

and $\left(\dfrac{\partial \ln x_A}{\partial n_A}\right)_{n_B} = \left[\dfrac{\partial\left(\dfrac{n_B}{n_A + n_B}\right)}{\partial n_A}\right]_{n_B} = -\dfrac{1}{n_A + n_B}$

$$\mu_A = \mu_A^* + RT\ln x_A + RT\left(1 - \frac{n_A}{n_A + n_B} - \frac{n_B}{n_A + n_B}\right)$$
$$+ C\left(\frac{n_A n_B + n_B^2 - n_A n_B}{(n_A + n_B)^2}\right)$$

$$\mu = \mu_A^* + RT\ln x_A + RT(0) + \frac{Cn_B^2}{(n_A + n_B)^2}$$

$$= \mu_A^* + RT\ln x_A + Cx_B^2$$

(b) Write

$$\mu_A = \mu_A^* + RT\ln x_A = \mu_A^* + RT\ln\gamma_A$$

By comparison $RT\ln\gamma_A = Cx_B^2$ or $\ln\gamma = \dfrac{Cx_B^2}{RT}$.

Thus $\gamma = e^{Cx_B^2/RT} = 1$ when $x_B \to 0$.

This corresponds to a pure A. In a very dilute solution of A in B we also expect $\gamma_A \to 1$.

In that case

$$\mu_A^{*\prime} = \lim_{x_A \to 0} (\mu_A - RT\ln x_A)$$

Substitution from above

$$\mu_A^{*\prime} = \lim_{x_B \to 1} (\mu_A^* + Cx_B^2) = \mu_A^* + C$$

Therefore

$$\mu_A = \mu_A^{*\prime} + RT\ln x_A + C(x_B^2 - 1) = \mu_A^* + RT\ln x_A + RT\ln\gamma_A$$

$$\ln\gamma_A = C(x_B^2 - 1)/RT = 0 \text{ when } x_B \to 1$$

CHAPTER 6: WORKED SOLUTIONS
PHASE EQUILIBRIUM

1. The compositions of the two phases at a particular temperature are: water saturated with nicotine, and nicotine saturated with water.

The number of degrees of freedom is

$$f = c - p + 2 = 2 - 2 + 2 = 2$$

This means that within the two-phase region, the temperature and weight % may be varied without changing the two-phase character of the region.

2. (a) For KCl and H_2O at the equilibrium pressure

$$f = c - p + 2 = 2 - 1 + 2 = 3$$

Since the equilibrium pressure is specified, this reduces the number of degrees of freedom to 2.

(b) Here, NaCl, KCl and H_2O are present. This is actually a three-component system since the solution contains Na^+, K^+, Cl^- and H_2O. The first three compositions are reduced to two independent ones by the electroneutrality condition. Therefore, $f = c - p + 2 = 3 - 1 + 2 = 4$, but with the restriction of constant pressure, the variance is reduced by 1, and is therefore 3.

(c) Ice, water and alcohol are only two components. Consequently, $f = c - p + 2 = 2 - 2 + 2 = 2$.

3. Aqueous sodium acetate is a two-component system even though the hydrolysis

$$\text{Acetate}^- + H_2O \rightleftharpoons OH^- + \text{HAc}$$

takes place, since the equilibrium constant defines the

concentration of OH^- and HAc if the concentration of sodium acetate is given.

4. Starting with pure $CaCO_3$ we have only one component present. When two of the three species are present the third species is also present, but because of the equilibrium $CaCO_3 \rightleftharpoons CaO + CO_2$ there are only two components.

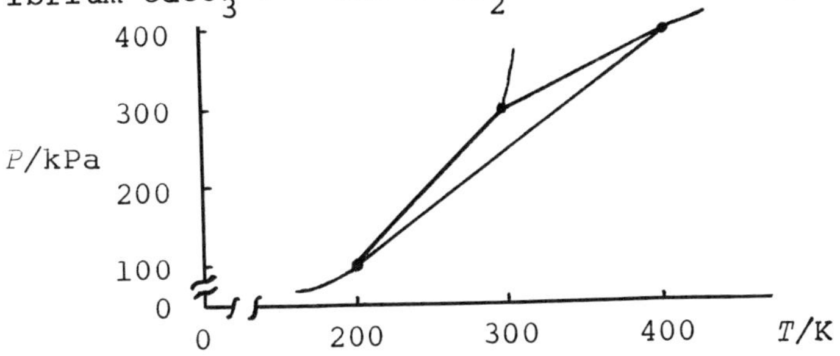

T/K	P/kPa	Phases Present
200	100	A, B, gas
300	300	A, B, liquid
400	400	B, liquid, gas

6. (a) Wt % of B = $\dfrac{99 \times 100}{33 + 99} = 75\%$

From the graph at 75% B, the first vapor appears at 60°C.

(b) The composition of the vapor is given by the intersection of the tie line at the vapor curve. In this case, the vapor has a composition of 88% B.

(c) The intersection of the liquid line at 65°C corresponds to 53% B in the liquid.

(d) Using the average composition of the distillates as the value midway between initial and final composition of the distillates, we have

$$\frac{1}{2}(88\% + 70\%) = 79\% \text{ B in distillate}$$

Let W_R = mass of residue; W_D = mass of distillate

$$W_R + W_D = 132$$

Then applying the condition that B is distributed through residue and distillate we have

mass of B in residue + mass of B in distillate = 99 g

$$0.53\ W_R + 0.79\ W_D = 99$$

$$0.53\ (132 - W_D) + 0.79\ W_D = 99$$

$$0.26\ W_D = 29$$

$$W_D = 111.5 \text{ g}; \quad W_R = 132 - 111.5 = 20.5 \text{ g}$$

In the distillate therefore, 79% of 111.5 g or 88.1 g is component B and 111.5 - 88.1 = 23.4 g is component A.

7. From Figure 6.14, the composition at 350 K at equilibrium between the single-phase water-rich region and the two-phase region is approximately 10% nicotine. For the equilibrium value on the nicotine-rich side the value is approximately 75% nicotine. Using the lever rule, we have

$$\frac{\text{Mass of water-rich layer}}{\text{Mass of nicotine-rich layer}} = \frac{75 - 40}{40 - 10} = \frac{35}{30} = 1.2$$

8. From Eq. 6.22, since $w = nM$, we write

$$\frac{w_A}{w_B} = \frac{n_A M_A}{n_B M_B} = \frac{P_A^* M_A}{P_B^* M_B} \text{ or } M_A = \frac{P_B^*}{P_A^*}\ \frac{w_A M_B}{w_B}$$

The vapor pressure of pure chlorobenzene is

56.434 - 43.102 = 13.332 kPa. Substitution gives

$$M_{\text{chlorobenzene}} = \frac{43.102}{13.334}\ (1.93)\ 18.02 = 112.4$$

The value obtained by addition of relative atomic masses is 112.6.

9. To determine the masses of the material distilled, the numerator of Eq. 6.22 is multiplied by M_A and the denominator by M_B. Since $w = nM$,

$$\frac{w_A}{w_B} = \frac{n_A M_A}{n_B M_B} = \frac{P_A^* M_A}{P_B^* M_B}$$

10. The vapor pressure of water at 372.4 K is 98.805 kPa. The vapor pressure of naphthalene is therefore 101.325 kPa - 98.805 kPa = 2.52 kPa. The molar mass of naphthalene is 128.19 g mol^{-1}. From the modified Eq. 6.22, we have

$$\frac{w_{H_2O}}{w_{\text{napth}}} = \frac{P^*_{H_2O}\ M_{H_2O}}{P^*_{\text{napth}}\ M_{\text{napth}}}$$

$$w_{H_2O} = \frac{98.805\ \text{kPa}\ (18.02\ \text{g})\ 1\ \text{kg}}{2.52\ \text{kPa}\ (128.19\ \text{g})} = 5.51\ \text{kg}$$

11. The vapor pressure of chlorobenzene is 56.434 - 43.102 = 13.332 kPa. From Problem 9,

$$\frac{\text{mass of chlorobenzene}}{\text{mass of water}} = \frac{13\ 332 \times 0.1125}{43\ 102 \times 0.01802} = 1.93$$

12. Using Eq. 6.19 and letting isoamyl alcohol be component 1 as depicted in Figure 6.2, we have

$$y_{IAA} = \frac{0.4 \times 2330}{0.4 \times 2330 \times 0.6 \times 7460} = 0.172$$

$$y_{IBA} = 1 - y_{IAA} = 1.000 - 0.172 = 0.828$$

13. Take the logarithms of both sides of $\rho = m/V$: $\ln \rho = \ln m - \ln V$. Partial differentiation with respect to T gives

$$\left(\frac{\partial \ln \rho}{\partial T}\right)_P = -\left(\frac{\partial \ln V}{\partial T}\right)_P = -\frac{1}{V}\left(\frac{\partial V}{\partial T}\right)_P = -\alpha$$

14. From the previous problem $\alpha = \frac{1}{V}\left(\frac{\partial V}{\partial T}\right)_T \approx \frac{1}{V}\left(\frac{\Delta V}{\Delta T}\right)_P$ for small changes in V and T. For an arbitrary quantity of water, say 1 gram exactly, the equation $V = m/\rho$ gives $V = 1.001769\ \text{cm}^3$ at 20°C and $V = 1.001982\ \text{cm}^3$ at 21°C and 1 atm.

Using these values we have

$$\alpha \approx \frac{1}{1.001769} \quad \frac{1.001982 - 1.001769}{1\ \text{K}} = \frac{0.000213}{1.001769\ \text{K}}$$

$$\approx 0.002126\ \text{K}^{-1} = 2.126 \times 10^{-4}\ \text{K}^{-1}$$

15. $\frac{\alpha}{\kappa} = \left(\frac{\partial P}{\partial T}\right)_V \approx \left(\frac{\Delta P}{\Delta T}\right)_V$ Therefore,

$$\Delta P \approx \frac{\alpha}{\kappa}\Delta T = \frac{2.85 \times 10^{-4}\ \text{K}^{-1}}{4.49 \times 10^{-5}\ \text{atm}^{-1}} \times 6\ \text{K} = 38.1\ \text{atm}$$

16. Application of the lever rule gives

$$\frac{\text{Mass of solid layer}}{\text{Mass of solid + liquid layer}} = \frac{\overline{se}}{\overline{be}} = \frac{0.13}{0.31} = 0.42$$

or 42% solid and 58% as liquid. The composition of the liquid is $x_{Si} = 0.31$.

17. The temperature at which solid solvent is in equilibrium with liquid solvent of mole fraction x_1 is given by Eq. 5.123,

$$\ln x_1 = \frac{H_{m,fus}}{R}\left(\frac{1}{T_{fus}} - \frac{1}{T}\right)$$

Values of x_1 and T determined from this equation for each component give the desired liquidus lines in the regions near large values of x_1. Several values are:

x_1	T/K	x_1	T/K
0.945	1650	0.969	1300
0.863	1600	0.924	1250
0.784	1550	0.879	1200
0.708	1500	0.783	1100
0.564	1400	0.681	1000

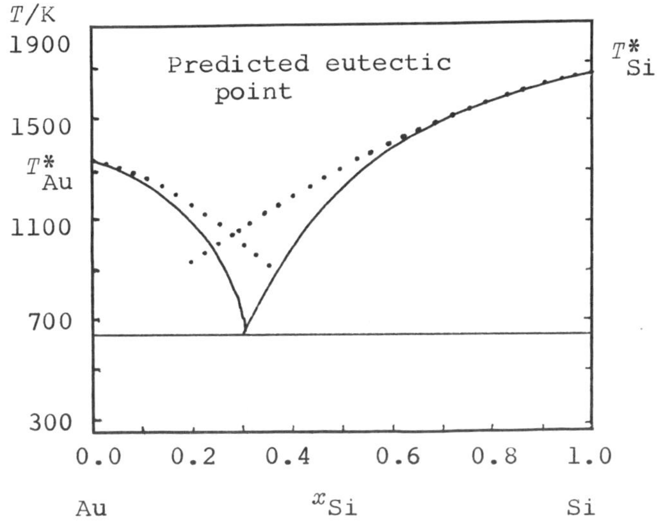

A plot is shown on which the points represent the data points and the solid curves are experimental curves of Figure 6.17. The dotted lines intersect at about $x_{Si} = 0.28$ compared to the actual $x_{Si} = 0.31$. However, the eutectic temperature is approximately 400 K too high.

18. (a) As water is added the saturated liquid of composition b would be in equilibrium with two solids A and B. At approximately 20% C when the composition crosses the line $\overline{bB}$, the solid A disappears and only solid B will be present in equilibrium with liquid of composition b.

(b) The two solid phases would not disappear until b is passed at approximately 50% liquid.

(c) Added water will cause dilution and solid salt will cease to exist.

19. (a) peritectic point

(b) eutectic point

(c) melting point

(d) incongruent melting

(e) phase transition

The figure for this occurs at the top of the next page.

20. Each halt corresponds to a line of three-phase equilibrium and each break to a boundary between a one- and a two-phase region. At 50% Y_2O_3, a compound is formed and may be written as $Fe_2O_3 \cdot Y_2O_3$ or $YFeO_3$.

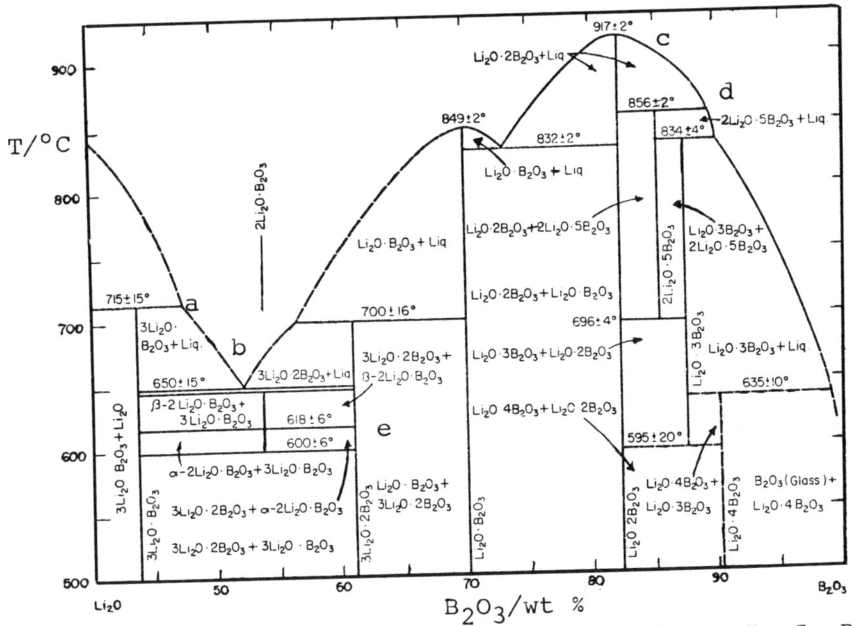

Copyright 1959, The American Ceramic Society, B. S. R. Sastry and F. A. Hummel, *J. Am. Ceram. Soc.*, 42 [5] 218 (1959).

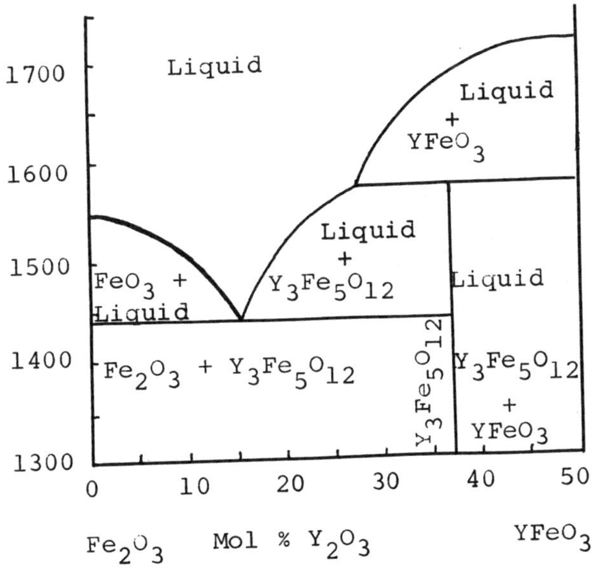

A compound unstable above 1575°C is indicated between 30 and 40% Y_2O_3. This might be taken to be $2Fe_2O_3 \cdot Y_2O_3$ at 33% Y_2O_3, but actually the formula is $Y_3Fe_5O_{12}$ corresponding to $3Y_2O_3 + 5Fe_2O_3$ at 37% Y_2O_3.

21. The diagrams are self-explanatory. The coexistence of the three phases is a clear indication of a peritectic-type diagram. A note of caution is in order here however. In the range 0 - 10 mol % Au at 1430°C to 1536°C still another phase, called δ, exists, and this would not be detected using only the compositions listed. One must be careful to use enough compositions to ensure that all the phases are identified. Also, the equilibrium between liquid and γ is not a simple curve and must be determined by careful experimentation.

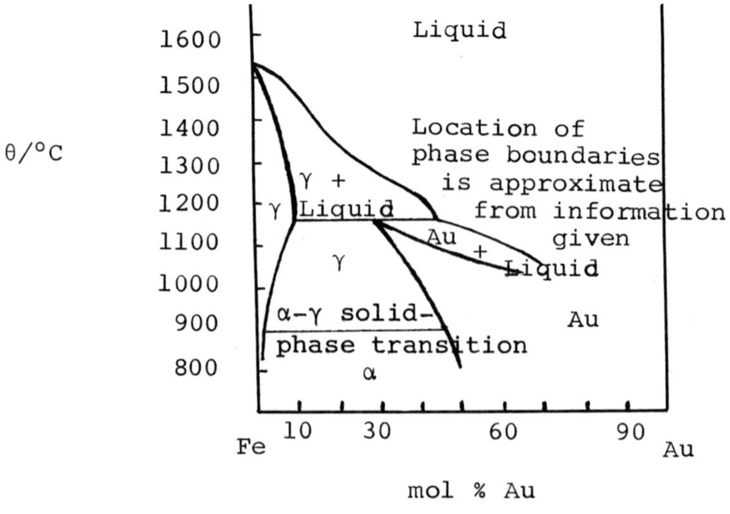

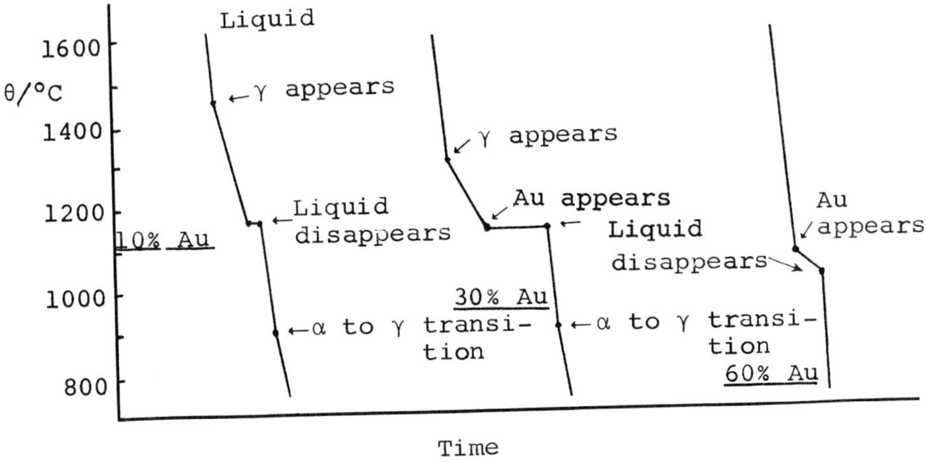
Liquid
1600
θ/°C
γ appears
1400
γ appears
1200
Liquid
disappears
Au appears
Au
appears
10% Au
Liquid
disappears
1000
30% Au
α to γ transition
α to γ transition
800
60% Au
Time

22.

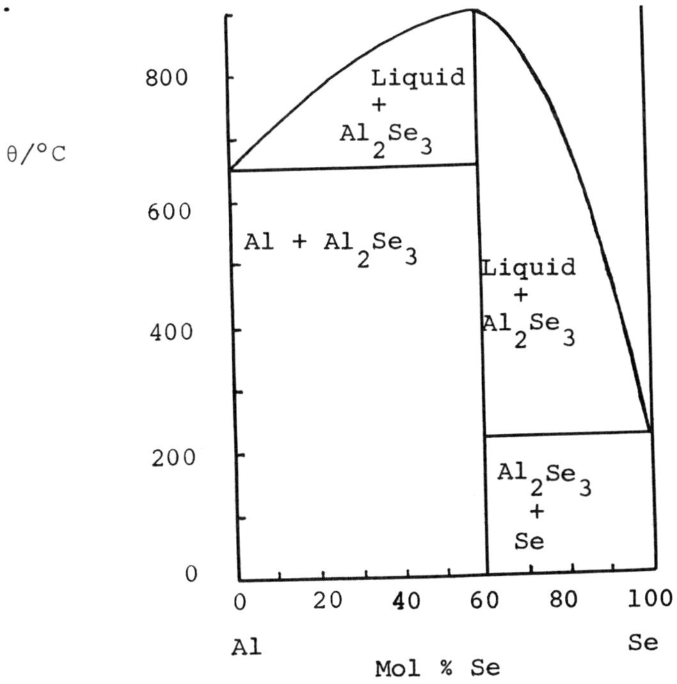
800
θ/°C
Liquid
+
Al2Se3
600
Al + Al2Se3
Liquid
+
Al2Se3
400
200
Al2Se3
+
Se
0
0 20 40 60 80 100
Al
Se
Mol % Se

23. The diagram is self-explanatory. The lower phase field of the α-phase is less than 1 percent.

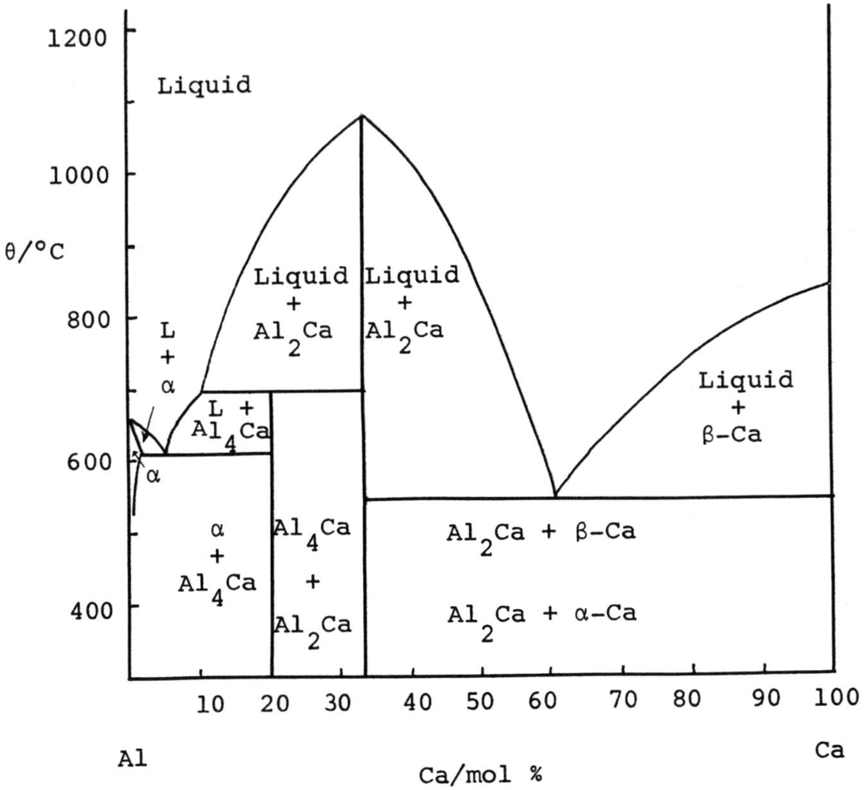

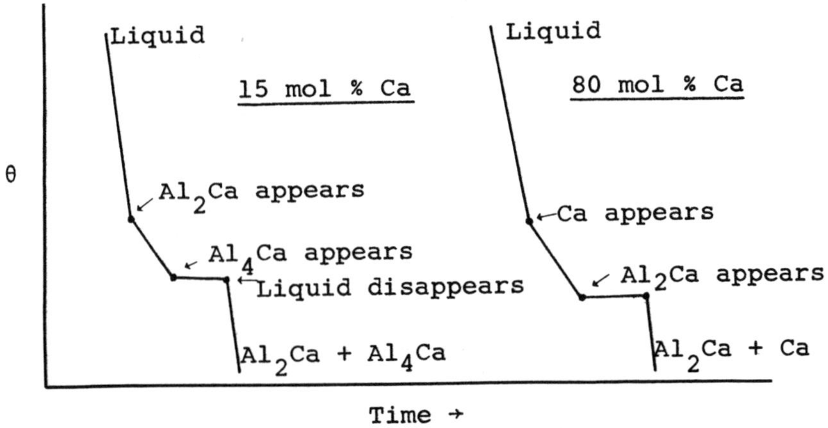

24.

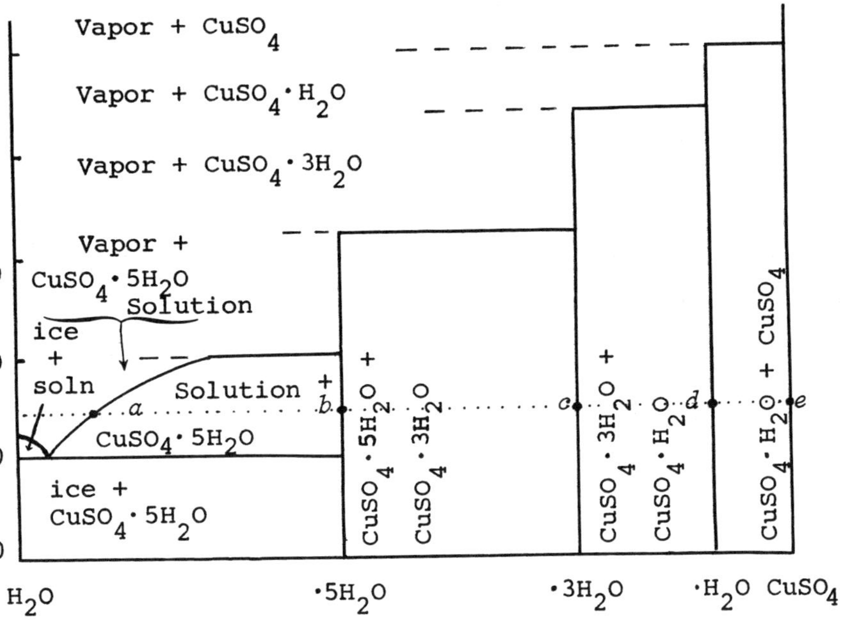
Vapor + CuSO4
310
Vapor + CuSO4·H2O
300
Vapor + CuSO4·3H2O
Vapor +
290
CuSO4·5H2O
Solution
T/K
ice
+
soln
280
Solution +
CuSO4·5H2O
a
b
c
d
e
270
ice +
CuSO4·5H2O
260
CuSO4·5H2O + CuSO4·3H2O
CuSO4·3H2O + CuSO4·H2O
CuSO4·H2O + CuSO4
H2O
·5H2O
·3H2O
·H2O CuSO4

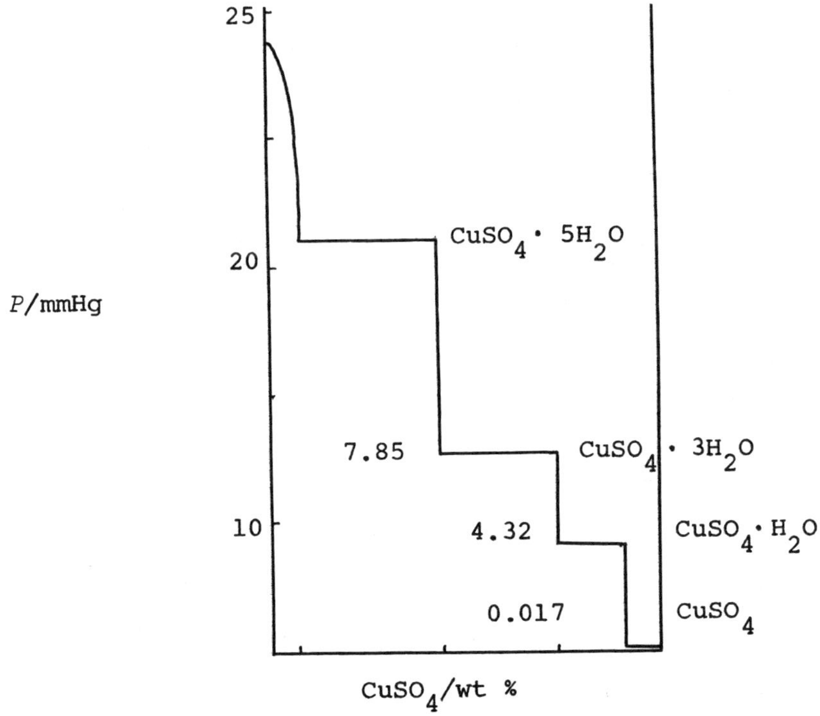
25
20
P/mmHg
10
CuSO4· 5H2O
7.85
CuSO4 · 3H2O
4.32
CuSO4·H2O
0.017
CuSO4
CuSO4/wt %

In the upper figure, the dilute $CuSO_4$ begins in a solution single-phase region. Pure $CuSO_4 \cdot 5H_2O$ precipitates out as the first phase boundary at *a* is crossed. Water continues to be removed as more $CuSO_4 \cdot 5H_2O$ precipitates until only pure $CuSO_4 \cdot 5H_2O$ is present at *b*. In the next two-phase region, $CuSO_4 \cdot 5H_2O$ dehydrates forming progressively more $CuSO_4 \cdot 3H_2O$ until all of the pentahydrate is gone at *c*. The process repeats, the trihydrate forming the monohydrate until only monohydrate is present at *d*. The monohydrate dehydrates until at *e* only pure $CuSO_4$ is present.

In the lower figure, the vapor pressure of water drops as the amount of $CuSO_4$ increases (Raoult's Law) until the solution is saturated with respect to the pentahydrate. The system is invariant since three phases, vapor, saturated solution, and solid $CuSO_4 \cdot 5H_2O$ are present at the constant temperature of 298.15 K. As the concentration of $CuSO_4$ increases (water is removed) the pressure remains constant until only $CuSO_4 \cdot 5H_2O$ is present. Removal of additional water causes some trihydrate to form and the pressure drops. Again the system is invariant; three phases are present, vapor, $CuSO_4 \cdot 5H_2O$ and $CuSO_4 \cdot 3H_2O$. The process is continued as before at the other stages.

25. The composition of the system is 30 mol % acetic acid, 50 mol % water, and 20 mol % toluene. The system point is practically on the p"-q" tie line, and there are therefore two liquids present. The ends of this line and thus the concentrations of the two liquids are approximately:

A: 95.5% toluene, 4% acetic acid, 0.5% H_2O and

B: 1% toluene, 37% acetic acid, and 62% H_2O.

The relative amounts of the two liquids are given by the lever rule: $\frac{15.4\ B}{3.8\ A}$, or $4B$ to $1A$.

26. The system CaO-MgO-SnO_2:

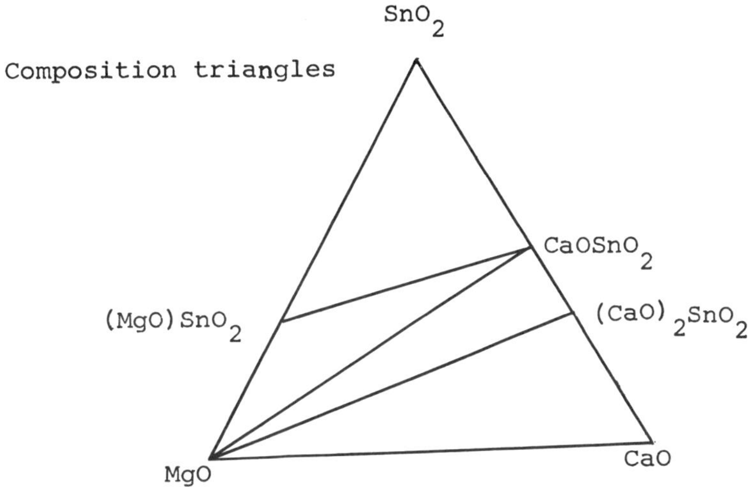

27.

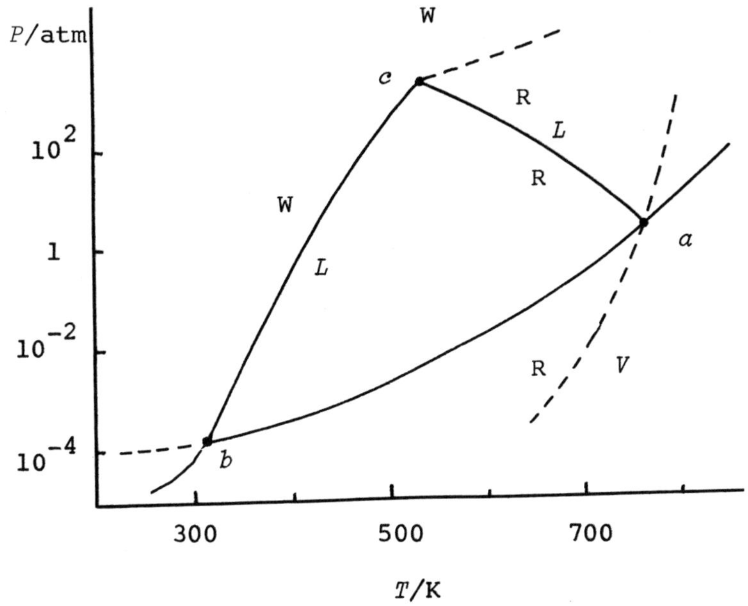

a) Stable triple point, R (red phosphorus, solid, liquid (*L*), vapor (*V*).

b) A metastable triple point, W (white phosphorus, solid), *L*, *V*. (The vapor pressure of the white form is greater than that of the red form.)

c) Stable triple point W, R, *L*. If we assume that a solid cannot be superheated, the triple point W, R, *L* is totally unstable since it probably lies above the melting point of the liquid.

28.

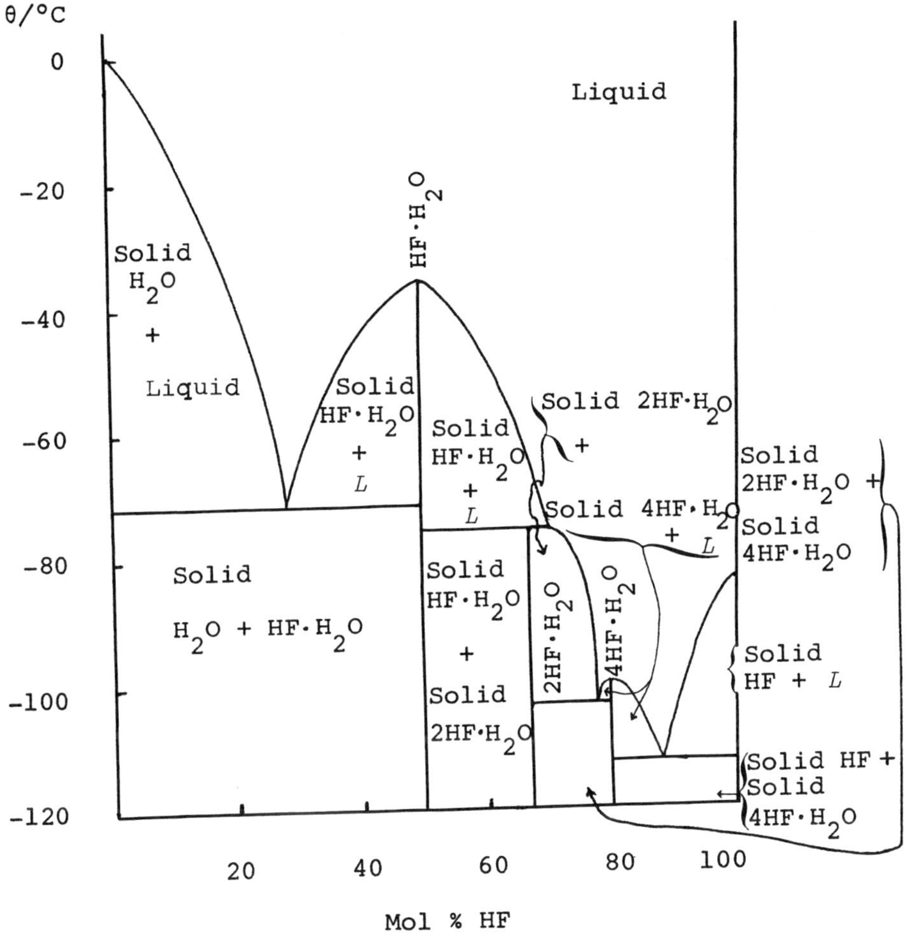

29. The 500 K equilibrium line probably contains an eutectic since the temperature is below both the melting point of AB_2 and B. An unstable compound is ruled out because such a reaction would require cooling halts at both 900 K and 500 K. Instead, a peritectic reaction shown is the simplest explanation.

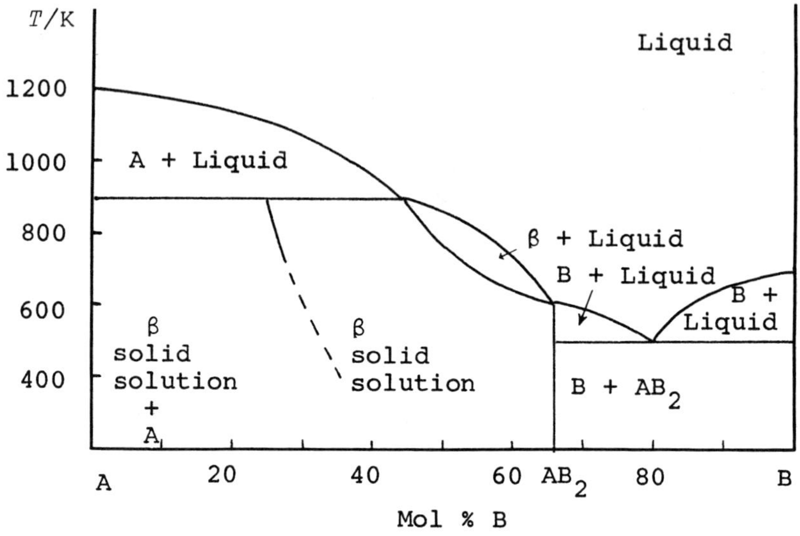
T/K
1200
1000
800
600
400
Liquid
A + Liquid
β + Liquid
B + Liquid
B + Liquid
β solid solution + A
β solid solution
B + AB2
A
20
40
60
AB2
80
B
Mol % B

30.

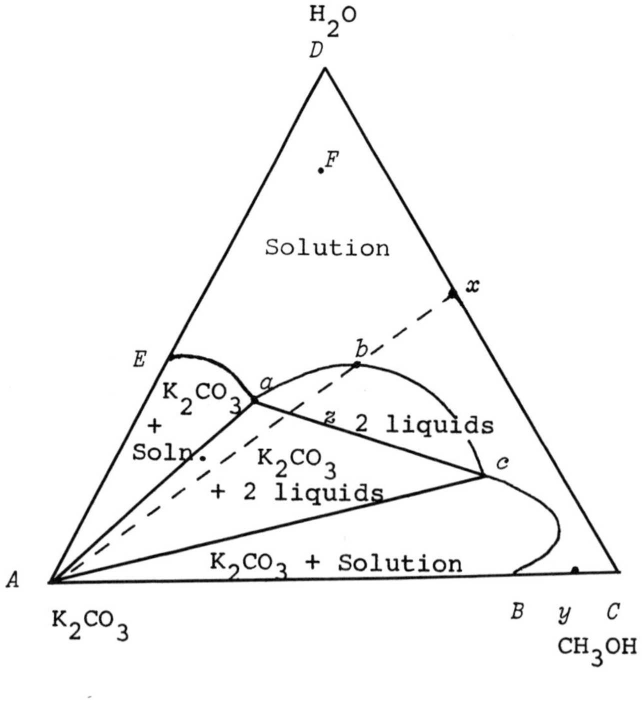
H2O
D
F
Solution
x
b
E
a
K2CO3 + Soln.
z
2 liquids
c
K2CO3 + 2 liquids
K2CO3 + Solution
A
K2CO3
B
y
C
CH3OH

a)

Region	System
AEa	K_2CO_3 in equilibrium with water-rich saturated solution
Aac	K_2CO_3 in equilibrium with conjugate liquids a and c
abc	two conjugate liquids joined by tie lines
AcB	K_2CO_3 in equilibrium with alcohol-rich saturated solution

b) The state of the system will move along a line joining *x* and *A*. Initially solution is formed; as more K_2CO_3 is added two layers *a* and *c* form and once beyond point *z*, K_2CO_3 ceases to dissolve so that solid K_2CO_3 and the two liquids *a* and *c* coexist.

c) As long as two liquids exist, liquid with composition in the region *AcB* is the alcohol-rich layer and may be separated from the water-rich layer using a separatory funnel.

d) When water is added to an unsaturated solution of K_2CO_3 in alcohol the state of the system moves along the line joining *y* and *D*. Some K_2CO_3 will precipitate as the state moves into the *ABc* region and then redissolves as it moves into the solution region again.

e) On evaporation of *F*, the system composition follows a line drawn from the water corner through *F* to the *Ac* line. At the first composition line, two liquids form and the compositions of the solutions move toward *a* and *c*. When the system composition reaches the *ac* line, K_2CO_3 begins to precipitate and is in equilibrium with the conjugate liquids *a* and *c*. Further reduction of water moves the ratio of liquid *a* to liquid *c* in favor of *c* until the line *Ac* is crossed, at which time solid K_2CO_3 is in equilibrium with a single solution.

CHAPTER 7: WORKED SOLUTIONS
SOLUTIONS OF ELECTROLYTES

1. 96 500 C deposits (63.5/2) g of copper; the quantity passed is therefore

$$\frac{96\ 500 \times 0.04 \times 2}{63.5}\ \text{C}$$

The current was passed for 3600 s; the current is therefore

$$\frac{96\ 500 \times 0.04 \times 2}{63.5 \times 3600}\ \text{A} = 0.03377\ \text{A} = 33.8\ \text{mA}$$

2. Quantity of electricity passed = $\frac{96\ 500 \times 0.00719}{107.9}$C

Current = $\frac{96\ 500 \times 0.00719}{107.9 \times 45 \times 60}$ A = 2.38×10^{-3} A = 2.38 mA

3.

$c/10^{-4}M$	$\frac{\Lambda}{\Omega^{-1}\text{cm}^2\text{mol}^{-1}}$	α	$1 - \alpha$	$\frac{K = c\alpha^2/(1-\alpha)}{10^{-3}\ M}$
625	53.1	.147	.853	1.583
312.5	72.4	.200	.800	1.563
156.3	96.8	.267	.733	1.520
78.1	127.7	.353	.647	1.504
39.1	164	.453	.547	1.467
19.6	205.8	.569	.431	1.472
9.8	249.2	.688	.312	1.487

The values are reasonably constant; average K = 1.51×10^{-3} mol dm^{-3}.

4. Λ_{AgCl} = 61.8 + 76.4 = 138.2 Ω^{-1} cm^2 mol^{-1}

Solubility = $\frac{1.26 \times 10^{-6}}{138.2}$ mol cm^{-3}

= 9.12×10^{-9} mol cm^{-3}

= 9.12×10^{-6} mol dm^{-3} = 9.12 μM

5. The increase in conductivity, $4.4 \times 10^{-4}\ \Omega^{-1}\ cm^{-1}$, is due to the $CaSO_4$ present; thus

$$\Lambda(\tfrac{1}{2}Ca^{2+} + \tfrac{1}{2}SO_4^{2-}) = \frac{4.4 \times 10^{-4}\ \Omega^{-1}\ cm^{-1}}{2c}$$

where c is the concentration of $CaSO_4$; $2c$ is the concentration of $\frac{1}{2}CaSO_4$. The value of $\lambda(\frac{1}{2}SO_4^{2-})$ is obtained from the conductivity of the Na_2SO_4 solution:

$$\Lambda(Na^+ + \tfrac{1}{2}SO_4^{2-}) = \frac{2.6 \times 10^{-4}\ \Omega^{-1}\ cm^{-1}}{2.0 \times 10^{-6}\ mol\ cm^{-3}}$$

(Note that since the concentration of Na_2SO_4 is 0.001 M, that of $\frac{1}{2}Na_2SO_4$ is 0.002 M). Thus

$$\Lambda(Na^+ + \tfrac{1}{2}SO_4^{2-}) = 130.0\ \Omega^{-1}\ cm^2\ mol^{-1}$$

Thus since $\lambda(Na^+) = 50.1\ \Omega^{-1}\ cm^2\ mol^{-1}$,

$$\lambda(\tfrac{1}{2}SO_4^{2-}) = 79.9\ \Omega^{-1}\ cm^2\ mol^{-1}$$

Then

$$\Lambda(\tfrac{1}{2}Ca^{2+} + \tfrac{1}{2}SO_4^{2-}) = 139.4\ \Omega^{-1}\ cm^2\ mol^{-1}$$

and

$$c = \frac{4.4 \times 10^{-4}\ \Omega^{-1}\ cm^{-1}}{2 \times 139.4\ \Omega^{-1}\ cm^2\ mol^{-1}} = 1.578 \times 10^{-3}\ mol\ dm^{-3}$$

Thus

$$c_{Ca^{2+}} = 1.578 \times 10^{-3}\ mol\ dm^{-3}$$

$$c_{SO_4^{2-}} = (1.0 \times 10^{-3} + 1.578 \times 10^{-3})\ mol\ dm^{-3}$$

$$= 2.578 \times 10^{-3}\ mol\ dm^{-3}$$

$$K_{sp} = 4.07 \times 10^{-6}\ mol^2\ dm^{-6}$$

6. $\Lambda(\mathrm{KCl}) = (73.5 + 76.4)\ \Omega^{-1}\ \mathrm{cm}^2\ \mathrm{mol}^{-1}$

The electrolytic conductivity at 0.01 *M* is

$\kappa(\mathrm{KCl}) = 149.9\ \Omega^{-1}\ \mathrm{cm}^2\ \mathrm{mol}^{-1} \times 10^{-5}\ \mathrm{mol\ cm}^{-3}$

$= 1.50 \times 10^{-3}\ \Omega^{-1}\ \mathrm{cm}^{-1}$

The electroytic conductivity of the ammonia solution is thus

$\kappa(\mathrm{NH_4OH}) = 1.50 \times 10^{-3} \times \frac{189}{2460} = 1.15 \times 10^{-4}\ \Omega^{-1}\ \mathrm{cm}^{-1}$

The molar conductivity of $NH_4^+ + OH^-$ is

$\Lambda(\mathrm{NH_4^+ + OH^-}) = (73.4 + 198.6)\ \Omega^{-1}\ \mathrm{cm}^2\ \mathrm{mol}^{-1}$

If $c = [\mathrm{NH_4^+}] = [\mathrm{OH^-}]$,

$272.0\ \Omega^{-1}\ \mathrm{cm}^2\ \mathrm{mol}^{-1} = \frac{1.15 \times 10^{-4}\ \Omega^{-1}\ \mathrm{cm}^{-1}}{c}$

$c = 4.23 \times 10^{-7}\ \mathrm{mol\ cm}^{-3} = 4.23 \times 10^{-4}\ \mathrm{mol\ dm}^{-3}$

The concentrations of NH_4OH, NH_4^+ and OH^- are thus

$$\mathrm{NH_4OH} \rightleftharpoons \mathrm{NH_4^+} + \mathrm{OH^-}$$

$0.01 - 4.23 \times 10^{-4}$ 4.23×10^{-4} $4.23 \times 10^{-4}\ \mathrm{mol\ dm}^{-3}$

$K_b = 1.87 \times 10^{-5}\ \mathrm{mol\ dm}^{-3}$

7. From Eq. 7.50,

thickness $\propto c^{-1/2}$

thickness $\propto \varepsilon^{1/2}$

Therefore,

(a) at 0.0001 *M*, thickness $= 0.964 \times \sqrt{1000} = 30.5$ nm

(b) at $\varepsilon = 38$, thickness $= 0.964 \times \sqrt{\frac{38}{78}} = 0.673$ nm

8. $\Lambda_{\frac{1}{2}\mathrm{Na_2SO_4}} = \Lambda_{\mathrm{NaCl}} + \Lambda_{\frac{1}{2}\mathrm{K_2SO_4}} - \Lambda_{\mathrm{KCl}}$

$$= 126.5 + 153.3 - 149.9 = 129.9\ \text{cm}^2\ \Omega^{-1}\ \text{mol}^{-1}$$

9. $\Lambda^\circ_{NH_4OH} = \Lambda^\circ_{NH_4Cl} - \lambda^\circ_{Cl^-} + \lambda^\circ_{OH^-}$

$$= 129.8 - 65.6 + 174.0 = 238.2\ \text{cm}^2\ \Omega^{-1}\ \text{mol}^{-1}$$

$$\alpha = \frac{9.6}{238.2} = .0403$$

10. Quantity of electricity = 2 h x 3600 s h^{-1} x 0.79 A

= 5688 C

Amount deposited = 5688/96 500 = 0.059 mol

Loss of LiCl in anode compartment = $\frac{0.793\ \text{g}}{42.39\ \text{g mol}^{-1}}$

= 0.0187 mol

Anode reaction: $Cl^- \rightarrow \frac{1}{2}Cl_2 + e^-$

0.059 mol Cl^- is removed by electrolysis

Net loss = 0.0187 mol Cl^-

0.059 - 0.0187 = 0.0403 mol Cl^- have migrated into the anode compartment

$$t_{Cl^-} = \frac{0.0403}{0.059} = 0.683$$

$$t_{Li^+} = 1 - 0.683 = 0.317$$

$$\lambda^\circ_{Li^+} = 0.317 \times 115.0 = 36.4\ \Omega^{-1}\ \text{cm}^2\ \text{mol}^{-1}$$

$$\lambda^\circ_{Cl^-} = 78.6\ \Omega^{-1}\ \text{cm}^2\ \text{mol}^{-1}$$

Then, from Eq. 7.64,

$$u_+ = 36.4/96\ 500 = 3.77 \times 10^{-4}\ \text{cm}^2\ \text{V}^{-1}\ \text{s}^{-1}$$

$$u_- = 78.6/96\ 500 = 8.15 \times 10^{-4}\ \text{cm}^2\ \text{V}^{-1}\ \text{s}^{-1}$$

11. Relative molecular mass of CdI_2 = 366.21

96 500 C deposits $\frac{1}{2}Cd^{2+}$ = 56.2 g of Cd^{2+}

$\therefore$ current passed is $\dfrac{0.034\ 62 \times 96\ 500}{56.2} = 59.45$ C

Anode compartment (152.64 g) originally contained

$\dfrac{7.545 \times 10^{-3} \times 152.64}{1000} = 1.1517 \times 10^{-3}$ mol.

It finally contains

$\dfrac{0.3718}{366.21} = 1.0153 \times 10^{-3}$ mol

Loss in anode compartment = 1.364×10^{-4} mol

96 500 C would have brought about a loss of

$\dfrac{1.364 \times 10^{-4} \times 96\ 500}{59.45} = 0.221$ mol of CdI_2

$= 0.442$ mol of $\frac{1}{2}$ CdI_2

$\therefore\ t_+ = 0.442;\ t_- = 0.558$

12. The individual ionic conductivities are:

$\lambda_+^\circ = 0.821 \times 426.16 = 349.9\ \Omega^{-1}\ cm^2\ mol^{-1}$

$\lambda_-^\circ = 0.179 \times 426.16 = 76.3\ \Omega^{-1}\ cm^2\ mol^{-1}$

Then, by Eq. 7.64, the ionic mobilities are

$u_+ = \dfrac{349.9\ \Omega^{-1}\ cm^2\ mol^{-1}}{96\ 500\ C} = 3.63 \times 10^{-3}\ cm^2\ V^{-1}\ s^{-1}$

$u_- = \dfrac{76.3\ \Omega^{-1}\ cm^2\ mol^{-1}}{96\ 500\ C} = 7.90 \times 10^{-4}\ cm^2\ V^{-1}\ s^{-1}$

13. The ionic mobilities are (Eq. 7.64)

$u_+ = \dfrac{50.1\ \Omega^{-1}\ cm^2\ mol^{-1}}{96\ 500\ C} = 5.19 \times 10^{-4}\ cm^2\ V^{-1}\ s^{-1}$

$u_- = \dfrac{76.4\ \Omega^{-1}\ cm^2\ mol^{-1}}{96\ 500\ C\ mol^{-1}} = 7.92 \times 10^{-4}\ cm^2\ V^{-1}\ s^{-1}$

The velocities in a gradient of 100 V cm^{-1} are thus

Na^+: 5.19×10^{-2} cm s^{-1}

Cl^-: 7.92×10^{-2} cm s^{-1}

14. The molar conductivity of LiCl is

$\Lambda = (38.6 + 76.4)\ \Omega^{-1}\ \text{cm}^2\ \text{mol}^{-1}$

The specific conductivity of a 0.01 M solution is this quantity multiplied by 10^{-4} mol cm^{-3}:

$\kappa = 115.0 \times 10^{-5}\ \Omega^{-1}\ \text{cm}^{-1}$

The resistance of a 1 cm length of tube is thus

$$R = \frac{1\ \text{cm}/5\ \text{cm}^2}{115.0 \times 10^{-5}\ \Omega^{-1}\ \text{cm}^{-1}} = 173.9\ \Omega$$

The potential required to produce a current of 1 A is

$$173.9\ \Omega \times 1\ \text{A} = 173.9\ \text{V}$$

The potential gradient is thus 173.9 V cm^{-1}.

The mobilities of the ions are (Eq. 7.64)

$$Li^+:\quad \frac{38.6\ \Omega^{-1}\ \text{cm}^2\ \text{mol}^{-1}}{96\ 500\ \text{C mol}^{-1}} = 4.00 \times 10^{-4}\ \text{cm}^2\ \text{V}^{-1}\ \text{s}^{-1}$$

$$Cl^-:\quad \frac{76.4\ \Omega^{-1}\ \text{cm}^2\ \text{mol}^{-1}}{96\ 500\ \text{C mol}^{-1}} = 7.92 \times 10^{-4}\ \text{cm}^2\ \text{V}^{-1}\ \text{s}^{-1}$$

The velocities are

Li^+: 0.070 cm s^{-1}; Cl^-: 0.138 cm s^{-1}

15. The work is given by $dw = Fdr$ where the force of attraction is

$$F = Q_1Q_2/r^2$$

Therefore

$$w = \int_{r_1}^{\infty} -\frac{Q_1Q_2}{4\pi\varepsilon_o r^2}\, dr = \frac{Q_1Q_2}{4\pi\varepsilon_o}\left(\frac{1}{\infty} - \frac{1}{r_1}\right)$$

(a) $\varepsilon_o = 8.854 \times 10^{-12}\ \text{C}^2\ \text{J}^{-1}\ \text{m}^{-1}$; $r_1 = 10^{-9}$ m

$$w = -\frac{(1.6 \times 10^{-19}\ \text{C})^2}{4\pi 8.85 \times 10^{-12}\ \text{C}^2\ \text{J}^{-1}\ \text{m}^{-1}}\left(\frac{1}{10^{-9}\ \text{m}}\right)$$

$= 2.30 \times 10^{-19}$ J

(b) $w = -2.30 \times 10^{-28} (1/\infty - 1/10^{-3}\text{ m})$

$= 2.30 \times 10^{-25}$ J

(c) $w = -2.30 \times 10^{-28} (1/0.10\text{ m} - 1/10^{-9}\text{ m})$

$= -2.30 \times 10^{-28} (10 - 10^{9}) = 2.30 \times 10^{-19}$ J

16. $NaCl$: $-239.7 - 167.4 = -407.1$ kJ mol^{-1}

$CaCl_2$: $-543.1 - 334.7 = -877.8$ kJ mol^{-1}

$ZnBr_2$: $-152.3 - (2 \times 120.9) = -394.1$ kJ mol^{-1}

17. H^+: -1051.4 kJ mol^{-1}

Na^+: $679.1 - 1051.4 = -372.3$ kJ mol^{-1}

Mg^{2+}: $274.1 - (2 \times 1051.4) = -1828.7$ kJ mol^{-1}

Al^{3+}: $-1346.4 - (3 \times 1051.4) = 4500.6$ kJ mol^{-1}

Cl^-: $-1407.1 + 1051.4 = -355.7$ kJ mol^{-1}

Br^-: $-1393.3 + 1051.4 = -341.9$ kJ mol^{-1}

18. KNO_3: $I = \frac{1}{2}(0.1 \times 1^2 + 0.1 \times 1^2) = 0.1\ M$

K_2SO_4: $I = \frac{1}{2}(0.2 \times 1^2 + 0.1 \times 2^2) = 0.3\ M$

$ZnSO_4$: $I = \frac{1}{2}(0.1 \times 2^2 + 0.1 \times 2^2) = 0.4\ M$

$ZnCl_2$: $I = \frac{1}{2}(0.1 \times 2^2 + 0.2 \times 1^2) = 0.3\ M$

$K_4Fe(CN)_6$: $I = \frac{1}{2}(0.4 \times 1^2 + 0.1 \times 4^4) = 1.0\ M$

19. Ionic strength of solution,

$I = \frac{1}{2}(0.4 \times 1^2 + 0.2 \times 2^2) = 0.6\ M$

$\log_{10} y_\pm = -z_+|z_-|0.51\sqrt{I}$

$= -2 \times 2 \times 0.51\sqrt{0.6}$

$= -2.04 \times .775 = -1.58$

$y_\pm = 0.026$

20. (a) $I = 1.274 \times 10^{-5}\ M$

$\log_{10} y_{\pm} = -0.51 \times (1.274 \times 10^{-5})^{1/2}$

$= -1.82 \times 10^{-3}$

$y_{\pm} = 0.996$

$K_s = y_{\pm} s^2 = (0.996 \times 1.274 \times 10^{-5})^2$

$= 1.609 \times 10^{-10}$

$\Delta G^{\circ} = -RT\ln K_s$

$= -8.314 \times 298.15 \ln 1.609 \times 10^{-10}$

$= 55.90\ \text{kJ mol}^{-1}$

(b) $I = \frac{1}{2}(0.01 + 0.005 \times 2^2) = 0.015\ M$

$\log_{10} y_{\pm} = -0.51\sqrt{0.015} = -0.0625$

$y_{\pm} = 0.866$

$$s = \frac{K_s^{1/2}}{y_{\pm}} = \frac{(1.609 \times 10^{-10})^{1/2}}{0.866} = 1.46 \times 10^{-5}\ M$$

22. The electrostatic contribution to Gibbs energy (Eq. 7.87) is, per mole of ions,

$$G^{\circ}_{\text{es}} = \frac{z^2 e^2 L}{8\pi\varepsilon_0 \varepsilon r}$$

$$= \frac{(1.602 \times 10^{-19})^2 \times 6.022 \times 10^{23}}{8\pi \times 8.854 \times 10^{-12} \times 0.133 \times 10^{-9}\ \varepsilon}\ \text{J mol}^{-1}$$

$$= \frac{5.22 \times 10^5}{\varepsilon}\ \text{J mol}^{-1}$$

In the membrane, $G^{\circ}_{\text{es}} = 130.5\ \text{kJ mol}^{-1}$; in water, $\Delta G^{\circ}_{\text{es}} = 6.7\ \text{kJ mol}^{-1}$

$\Delta G^{\circ}_{\text{es}}$ (water → membrane) $= 123.8\ \text{kJ mol}^{-1}$

23. $\Lambda^{\circ}\ (\frac{1}{2}\ CaF_2) = 51.0 + 47.0 = 98.0\ \Omega^{-1}\ \text{cm}^2\ \text{mol}^{-1}$

Observed κ due to salt $= 3.86 \times 10^{-5} - 1.5 \times 10^{-6}$

$$= 3.71 \times 10^{-5}\ \Omega^{-1}\ cm^{-1}$$

$$\text{solubility} = \frac{3.70 \times 10^{-5}}{98.}\ mol\ cm^{-3}$$

$$= 3.786 \times 10^{-4}\ mol\ dm^{-3}\ (\text{of } \tfrac{1}{2}\ CaF_2)$$

1 mol of $\frac{1}{2}$ CaF_2 has a mass of 20.04 + 19.00 = 39.04 g

Solubility = 0.0147 g dm^{-3}

$$\text{Solubility product} = [Ca^{2+}][F^-]^2$$

$$= (0.5 \times 3.786 \times 10^{-4}) \times (3.786 \times 10^{-4})^2$$

$$= 2.71 \times 10^{-11}\ mol^3\ dm^{-9}.$$

24. <u>Palmitate side</u> — <u>Other side</u>

Initial concentrations:

Palmitate side	Other side
$[Na^+] = 0.1\ M$	$[Na^+] = 0.2\ M$
$[P^-] = 0.1\ M$	$[Cl^-] = 0.2\ M$

Final concentrations:

Palmitate side	Other side
$[Na^+] = (0.1 + x)M$	$[Na^+] = (0.2 - x)M$
$[P^-] = 0.1\ M$	$[Cl^-] = (0.2 - x)M$
$[Cl^-] = x\ M$	

Then

$$(0.2 - x)^2 = (0.1 + x)$$

$$0.04 - 0.4\ x + x^2 = x^2 + 0.1\ x$$

$$x = \frac{0.04}{0.5} = 0.08$$

Final concentrations are thus, on the palmitate side,

$[Na^+] = 0.18\ M$; $[Cl^-] = 0.08\ M$

On the other side, $[Na^+] = [Cl^-] = 0.12\ M$

25. $x\ M\ CuSO_4$: $I = \frac{1}{2}(2^2 + 2^2)x = 4\ x\ M$

$I = 0.1\ M$ if $x = 0.025$

$x\ M\ Ni(NO_3)_2$: $I = \frac{1}{2}(2^2 + 2)x = 3\ x$

$I = 0.1\ M$ if $x = 0.033$

$x\ M\ Al_2(SO_4)_3$: $I = \frac{1}{2}(2 \times 3^2 + 3 \times 2^2)x = 15\ x\ M$

$I = 0.1\ M$ if $x = 0.00667$

$x\ M\ Na_3PO_4$: $I = \frac{1}{2}(3 + 3^2)x = 6\ x\ M$

$I = 0.1\ M$ if $x = 0.0167$

26. (a) First neglect the effect of activity coefficients: if s is the solubility

$s(2s)^2 = 4.0 \times 10^{-9}$ mol^3 dm^{-9}

$s = 1.0 \times 10^{-3}$ mol dm^{-3}

The ionic strength is

$\frac{1}{2}(1 \times 2^2 + 2 \times 1)1.0 \times 10^{-3} = 3.0 \times 10^{-3}$ mol dm^{-3}

By the Debye-Hückel limiting law

$\log_{10}y_\pm = -0.51 \times 2 \times \sqrt{3.0 \times 10^{-3}} = -0.056$

$y_\pm = 0.88$

If the solubility is s the activities of the ions are

Pb^{2+}: y_+s; F^-: $2\ y_-s$

Then

$(y_+s)(2\ y_-s)^2 = 4.0 \times 10^{-9}$ mol^3 dm^{-3}

$y_+y_-^2\ 4s^3 = 4.0 \times 10^{-9}$ mol^3 dm^{-9}

$y_\pm^3\ 4s^3 = 4.0 \times 10^{-9}$ mol^3 dm^{-9} (from Eq. 7.105)

Thus

$$s^3 = \frac{4.0 \times 10^{-9} \text{ mol}^3 \text{ dm}^{-9}}{(0.88)^3 \times 4}$$

$$s = 1.14 \times 10^{-3} \text{ mol dm}^{-3}$$

We could proceed to further approximations as necessary.

(b) In 0.01 *M* NaF the ionic strength is essentially 0.01 mol dm^{-3} and

$$\log_{10} y_{\pm} = -2 \times 0.51 \times \sqrt{0.01} = -0.102$$

$$y_{\pm} = 0.791$$

If s is the solubility

$$s = [Pb^{2+}];\ [F^-] = 0.01 \text{ mol dm}^{-3}$$

Then

$$s y_+ \times (0.01\ y_-)^2 = 4.0 \times 10^{-9} \text{ mol dm}^{-3}$$

$$y_+ y_-^2\ s \times 0.0001 = 4.0 \times 10^{-9} \text{ mol dm}^{-3}$$

$$y_{\pm}^3\ s \times 0.0001 = 4.0 \times 10^{-9} \text{ mol dm}^{-3}$$

$$s = \frac{4.0 \times 10^{-9} \text{ mol dm}^{-3}}{0.0001(0.791)^3} = 8.08 \times 10^{-5} \text{ mol dm}^{-3}$$

27. We proceed by successive approximations, first taking the activity coefficients to be unity. Then, if s is the solubility

$$s^2 = 4.0 \times 10^{-3} \text{ mol}^2 \text{ dm}^{-6}$$

$$s = 0.0632 \text{ mol dm}^{-3}$$

This is the ionic strength, and thus

$$\log_{10} y_{\pm} = -0.51\sqrt{0.0632} = -0.128$$

$y_\pm = 0.74$

To a second approximation

$$y_\pm s^2 = (0.74)^2 s^2 = 4.0 \times 10^{-3}\ \text{mol}^2\ \text{dm}^{-6}$$

$$s = 0.085\ \text{mol dm}^{-3}$$

To a third approximation

$$\log_{10} y_\pm = -0.51\sqrt{0.085};\ y_\pm = 0.71$$

$$(0.71)^2 s^2 = 4.0 \times 10^{-3}\ \text{mol}^2\ \text{dm}^{-6}$$

$$s = 0.089\ \text{mol dm}^{-3}$$

To a fourth approximation

$$\log_{10} y_\pm = -0.51\sqrt{0.089};\ y_\pm = 0.704$$

$$(0.704)^2 s^2 = 4.0 \times 10^{-3}$$

$$s = 0.090\ \text{mol dm}^{-3}$$

28. Equation 7.20 can be rearranged to

$$c\Lambda^2 = K\Lambda_0^2 - K\Lambda_0\Lambda$$

$c\Lambda^2$ could therefore be plotted against Λ. Alternatively, since

$$c\Lambda = K\Lambda_0^2 \cdot \frac{1}{\Lambda} - K\Lambda_0,$$

$c\Lambda$ can be plotted against $1/\Lambda$. The slope and intercepts are as shown below:

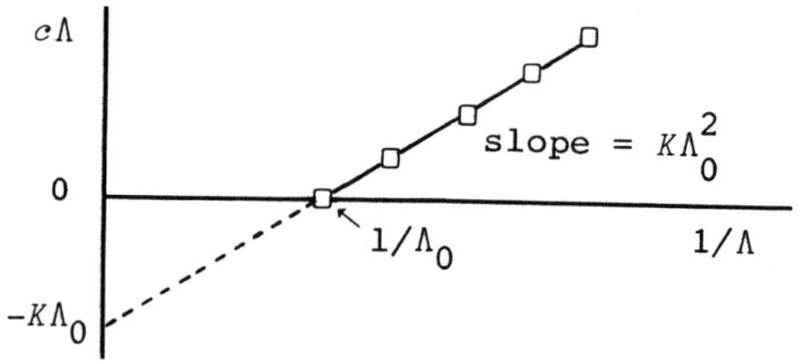

Λ values are obtained by use of Eq. 7.9; for the lowest concentration, 1.566×10^{-4} mol dm^{-3},

$$\Lambda = \frac{1.788 \times 10^{-6}\ \Omega^{-1}\ \text{cm}^{-1} \times 1000\ \text{dm}^{-3}\ \text{cm}^{3}}{1.566 \times 10^{-4}\ \text{mol dm}^{-3}}$$

$$= 11.4\ \Omega^{-1}\ \text{cm}^{2}\ \text{mol}^{-1}$$

Similarly for the other concentrations:

$c/10^{-4}$ mol dm^{-3}	1.566	2.600	6.219	10.441
Λ/Ω^{-1} cm^2 mol^{-1}	11.4	9.30	6.45	5.11
$c\Lambda/10^{-6}$ Ω^{-1} cm^{-1} mol^{-1}	1.785	2.418	4.011	5.335
$1/(\Lambda/\Omega^{-1}$ cm^2 mol$^{-1})$	0.0877	0.1075	0.155	0.196

In a plot of $c\Lambda$ against $1/\Lambda$, the intercepts are

$$-K\Lambda_0 = -1.15 \times 10^{-6}\ \Omega^{-1}\ \text{cm}^{-1}\ \text{mol}^{-1}$$

$$1/\Lambda_0 = 0.035\ \Omega^{-1}\ \text{cm}^{2}\ \text{mol}^{-1};\ \Lambda_0 = 30\ \Omega^{-1}\ \text{cm}^{2}\ \text{mol}^{-1}$$

$$K = 4.0 \times 10^{-8}\ \text{mol cm}^{-3} = 4.0 \times 10^{-5}\ \text{mol dm}^{-3}$$

29. The exponential is shown as curve a, $4\pi r^2$ as curve b, and their product as curve c, on the next page.

With $z_c = 1$ and $z_i = -1$ the function to be differentiated is

$$f = e^{e^2/4\pi\varepsilon_0\varepsilon r\mathbf{k}T}\ 4\pi r^2$$

Differentiation gives

$$\frac{df}{dr} = 8\pi r e^{e^2/4\pi\varepsilon_0\varepsilon r\mathbf{k}T} - 4\pi r^2 \cdot \frac{e^2}{4\pi\varepsilon_0\varepsilon r^2\mathbf{k}T}\ e^{e^2/4\pi\varepsilon_0\varepsilon r\mathbf{k}T}$$

Setting this equal to zero leads to

$$r^* = \frac{e^2}{8\pi\varepsilon_0\varepsilon\mathbf{k}T}$$

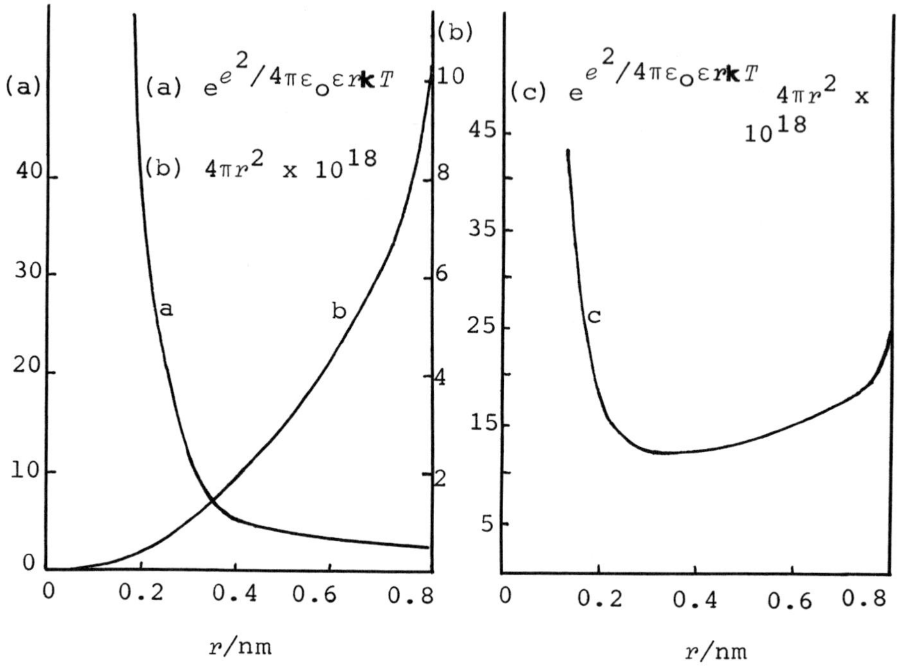

The value of this at 25.0°C, with $\varepsilon = 78.3$, is

$$3.58 \times 10^{-10}\ \text{m} = 0.358\ \text{nm}$$

With $z_c = 1$ the potential energy for two univalent ions, from Eq. 7.47, is

$$E_p = \frac{e^2}{4\pi\varepsilon_0\varepsilon r}$$

Introduction of the expression for r^* gives

$$E_p = 2\mathbf{k}T$$

At 25.0°C, $E_p = 8.23 \times 10^{-21}\ \text{J} = 4.96\ \text{kJ mol}^{-1}$

30. For Problem 22 it was found that

$$G^{\circ}_{es} = \frac{5.22 \times 10^5}{\varepsilon}\ \text{J mol}^{-1}$$

For the transfer from water (ε_1) to lipid (ε_2)

$$\Delta G^{\circ}_{es}/\text{J mol}^{-1} = 5.22 \times 10^5 \left(\frac{1}{\varepsilon_2} - \frac{1}{\varepsilon_1}\right)$$

$$\Delta S^\circ_{es} = -\left(\frac{\partial \Delta G^\circ_{es}}{\partial T}\right)_P \qquad \text{(from Eq. 3.119)}$$

Since ε_2 is temperature independent this leads to

$$\Delta S^\circ_{es}/\text{J K}^{-1}\ \text{mol}^{-1} = 5.22 \times 10^5 \frac{\partial}{\partial T}\left(\frac{1}{\varepsilon_1}\right)$$

$$= -5.22 \times 10^5 \frac{1}{\varepsilon_1^2} \frac{\partial \varepsilon}{\partial T}$$

$$= -5.22 \times 10^{-5} \frac{1}{\varepsilon_1} \cdot \frac{\partial \ln \varepsilon}{\partial \text{T}}$$

$$= \frac{5.22 \times 10^{-5} \times 0.0046}{78} = 31\ \text{J K}^{-1}\ \text{mol}^{-1}$$

The entropy increases because of the release of bound water molecules when the K^+ ions pass into the lipid.

31. (a) At infinite dilution the work of charging of an ion is given directly by (Eq. 7.86)

$$w_{rev} = \frac{z^2 e^2}{8\pi\varepsilon_o \varepsilon r}$$

For 1 mol of Na^+

$$w_{rev} = \frac{(1.602 \times 10^{-19}\ \text{C})^2\ 6.022 \times 10^{23}\ \text{mol}^{-1}}{8\pi \times 8.854 \times 10^{-12}\ \text{C}^2\ \text{N}^{-1}\ \text{m}^{-2} \times 78} \times \frac{1}{95 \times 10^{-12}\ \text{m}}$$

$$= 9373\ \text{J mol}^{-1}$$

For 1 mol of Cl^-,

$$w_{rev} = 4920\ \text{J mol}^{-1}$$

For 1 mol of Na^+Cl^- at infinite dilution

$$w_{rev} = 14\ 293\ \text{J mol}^{-1} = 14.3\ \text{kJ mol}^{-1}$$

(b) These values are reduced when the electrolyte is at a higher concentration, the work of charging the ionic atmosphere being negative and equal to $kT\ln y_i$. Thus for 1 mol of Na^+ ions, of activity coefficient y_+, the work of charging the atmosphere is

$$RT\ln y_+$$

Similarly for the chloride ion the work per mole is

$$RT\ln y_-$$

For 1 mol of Na^+Cl^-

$$w_{rev}(\text{atm}) = RT(\ln y_+ + \ln y_-)$$

$$= RT\ln y_+y_- = 2RT\ln y_\pm$$

If $y_\pm = 0.70$

$$w_{rev}(\text{atm}) = 2(8.314 \times 298.15\ \text{J mol}^{-1})\ \ln 0.70$$

$$= -1768\ \text{J mol}^{-1}$$

The net work of charging is thus

$$w_{rev} = 14\ 293 - 1768 = 12\ 525\ \text{J mol}^{-1}$$

$$= 12.5\ \text{kJ mol}^{-1}$$

CHAPTER 8: WORKED SOLUTIONS
ELECTROCHEMICAL CELLS

1. (a) $H_2 \rightarrow 2H^+ + 2e^-$

$Cl_2 + 2e^- \rightarrow 2Cl^-$

$H_2 + Cl_2 \rightarrow 2H^+ + 2Cl^-;\ z = 2$

$$E = E° - \frac{RT}{2F} \ln\left(a_H^2 + a_{Cl^-}^2\right)^u$$

(b) $2Hg(\ell) + 2Cl^- \rightarrow Hg_2Cl_2 + 2e^-$

$2H^+ + 2e^- \rightarrow H_2$

$2Hg + 2H^+ + 2Cl^- \rightarrow Hg_2Cl_2 + H_2;\ z = 2$

$$E = E° + \frac{RT}{2F} \ln\left(a_H^2 + a_{Cl^-}^2\right)^u$$

(c) $Ag + Cl^- \rightarrow AgCl(s) + e^-$

$2e^- + Hg_2Cl_2(s) \rightarrow 2Hg + 2Cl^-$

$2Ag(s) + Hg_2Cl_2(s) \rightarrow 2AgCl(s) + 2Hg(s)$

$E = E°$ (no concentration dependence)

(d) $\frac{1}{2}H_2(g) \rightarrow H^+ + e^-$

$AuI(s) + e^- \rightarrow Au(s) + I^-$

$AuI(s) + \frac{1}{2}H_2(g) \rightarrow Au(s) + H^+ + I^+;\ z = 1$

$$E = E° - \frac{RT}{F} \ln\ (a_{H^+}a_{I^-})^u$$

(e) $Ag(s) + Cl^-(a_1) \rightarrow AgCl(s) + e^-$

$AgCl(s) + e^- \rightarrow Ag(s) + Cl^-(a_2)$

$Cl^-(a_1) \rightarrow Cl^-(a_2);\ z = 1$

$$E = \frac{RT}{F} \ln \frac{a_1}{a_2}$$

2. (a) AH_2 is oxidized by B

(b) 0.44 V

(c) None from information given

3. The ΔG° values for the two reactions are

$Cr^{3+} + 3e^- \rightarrow Cr \quad \Delta G_1^\circ = -zE_1^\circ F = -3 \times -0.74 \times 96\ 500\ \text{J mol}^{-1}$

$Cr^{3+} + e^- \rightarrow Cr^{2+} \quad \Delta G_2^\circ = -zE_2^\circ F = -1 \times -0.41 \times 96\ 500\ \text{J mol}^{-1}$

The reaction $Cr^{2+} + 2e^- \rightarrow Cr$ is obtained by subtracting reaction (2) from reaction (1) and the ΔG° value for $Cr^{2+} + 2e^- \rightarrow Cr$ is obtained by subtracting ΔG_2° from ΔG_1°:

$\Delta G^\circ = -3 \times -0.74 \times 96\ 500 - (-1 \times -0.41 \times 96\ 500)\ \text{J mol}^{-1} = 1.81 \times 96\ 500\ \text{J mol}^{-1}$

Since $Cr^{2+} + 2e^- \rightarrow Cr$ involves 2 electrons and since $\Delta G = -zE^\circ F$, it follows that

$$1.81\ F = -2(E^\circ/\text{V})F$$

or

$$E^\circ = -0.905\ \text{V}$$

4. $E^\circ = 0.771 - 0.5355 = 0.2355$ V

$$\log_{10}K_c^u = \frac{2 \times 0.2355}{0.059\ 16} = 7.961$$

$K_c = 9.15 \times 10^7\ \text{dm}^6\ \text{mol}^{-2}$

5. The E° for the process

$$Sn + Fe^{2+} \rightleftharpoons Sn^{2+} + Fe$$

is $-0.4402 - (-0.136) = -0.304$ V

Since $z = 2$ we have

$$-0.304 = \frac{0.059\ 16}{2} \log_{10}K_c^u$$

$$K_c = 5.2 \times 10^{-11}$$

6. $\Delta G^\circ = -z \times 96\ 487 \times 0.25$ J; $z = 1$

$= -24\ 100$ J mol^{-1} $= -24.1$ kJ mol^{-1}

7. From Table 8.1 the relevant values are

(1) $O_2 + 4\,H^+ + 4e^- \rightarrow 2H_2O$ $\quad E^\circ = 1.23$ V

(2) $2H^+ + 2e^- \rightarrow H_2$ $\quad E^\circ = 0$

Subtraction of (2) from $\frac{1}{2}$(1) gives the required equation

$H_2 + \frac{1}{2}O_2 \rightleftharpoons H_2O$

with $E^\circ = 1.23$ V and $z = 2$. Then,

$\Delta G^\circ = -2 \times 96\ 487 \times 1.23$ J mol^{-1}

$= -237\ 400$ J mol^{-1} $= -237.4$ kJ mol^{-1}

8. From Table 8.1

(1) $Cu^{2+} + 2e^- \rightarrow Cu$ $\quad E_1^\circ = 0.337$ V

(2) $Cu^{2+} + e^- \rightarrow Cu^+$ $\quad E_2^\circ = 0.153$ V

If we subtract 2 x (2) from (1) we obtain

$2Cu^+ \rightarrow Cu^{2+} + Cu \quad E^\circ = E_1^\circ - E_2^\circ = 0.184$ V; $z = 2$

$\therefore \log_{10} K^u = \dfrac{2 \times 0.184}{0.059\ 16} = 6.22$

$K = 1.66 \times 10^6$ dm^3 mol^{-1}

If Cu_2O is dissolved in dilute H_2SO_4, half will form Cu^{2+} and half Cu.

9. $E = \dfrac{RT}{zF} \ln \dfrac{m_1}{m_2}$

$= 0.059\ 16 \log_{10} 2 = 0.0178$ V

10. The process is

pyruvate$^-$ + $2H^+ + 2e^- \rightarrow$ lactate$^-$

and the Nernst equation is

$$E' = E^\circ - \frac{RT}{2F} \ln \frac{[\text{lactate}^-]}{[\text{pyruvate}^-]}$$

Then

$$E'/\text{V} = -0.185 - \frac{0.059\ 16}{2} \log_{10}\left(\frac{10}{90}\right)$$

$$E' = -0.157\ \text{V}$$

11. (a) From Table 8.1

(1) $Fe^{3+} + e^- \rightarrow Fe^{2+}$ $E_1^\circ = 0.771$ V

(2) $Fe^{2+} + 2e^- \rightarrow Fe$ $E_2^\circ = -0.4402$ V

The corresponding ΔG° values are:

(1) $\Delta G_1^\circ = -0.771 \times 96\ 487 = -74\ 390\ \text{J mol}^{-1}$

(2) $\Delta G_2^\circ = -2\ (-0.4402) \times 96\ 487 = 84\ 940\ \text{J mol}^{-1}$

The required reaction is obtained by adding (1) and (2):

$Fe^{3+} + 3e^- \rightarrow Fe$ $\Delta G^\circ = 10\ 560\ \text{J mol}^{-1}$

Then, since $z = 3$,

$$E^\circ = -\frac{10\ 560}{3 \times 96\ 487} = -0.036\ \text{V}$$

(b) The electrode reactions are

(1) $Sn^{2+} \rightarrow Sn^{4+} + 2e^-$ $E_1^\circ = -0.15$ V

(2) $3e^- + Fe^{3+} \rightarrow Fe$ $E_1^\circ = -0.036$ V

The cell reaction is

$3Sn^{2+} + 2Fe^{3+} \rightarrow 3Sn^{4+} + 2Fe$ $E^\circ = -0.186$ V; $z = 6$

From the Nernst equation, Eq. 8.13,

$$E = -0.186 - \frac{0.059\ 16}{6} \log_{10} \frac{(0.01)^3}{(0.1)^3\ (0.5)^2}$$

$= -0.186 + 0.024 = -0.162$ V

12. For the reaction

$I_2 + I^- \rightarrow I_3^- \qquad E^\circ = -0.0010$ V and $z = 2$

$$K_c = \frac{[I_3^-]}{[I^-]} = \frac{[I_3^-]}{0.5 \text{ mol dm}^{-3}}$$

$$-0.0010 = \frac{0.059\ 16}{2} \log_{10} \frac{[I_3^-]}{0.5 \text{ mol dm}^{-3}}$$

$$\log_{10} \frac{[I_3^-]^u}{0.5} = -0.0338$$

$$\frac{[I_3^-]^u}{0.5} = 0.925$$

$[I_3^-] = 0.462$ *M*

13. $$\Delta\Phi/\text{V} = \frac{8.314 \times 298.15}{96\ 487} \ln \frac{0.18}{0.12}$$

$\Delta\Phi = 0.0104$ V $= 10.4$ mV

14. Reaction at the right-hand electrode:

$e^- + H^+ \ (0.2\ m) \rightarrow \frac{1}{2}H_2 \ (10 \text{ atm})$

At the left-hand electrode:

$\frac{1}{2}H_2 \ (1 \text{ atm}) \rightarrow e^- + H^+ \ (0.1\ m)$

Overall reaction:

$H^+(0.2\ m) + \frac{1}{2}H_2(1 \text{ atm}) \rightarrow \frac{1}{2}H_2(10 \text{ atm}) + H^+(0.1\ m)$;

$z = 1$

Cell emf:

$$E = \frac{RT}{F} \ln \frac{0.2 \times (1 \text{ atm})^{1/2}}{0.1 \times (10 \text{ atm})^{1/2}}$$

$$E/\text{V} = 0.059\ 16 \log_{10}\frac{2}{\sqrt{10}} = -0.0118 \text{ V}$$

$E = -11.8$ mV

15. The capacitance of the cell membrane (see Eq. 8.20) is

$$C = \frac{8.854 \times 10^{-12} \times 3 \times 10^{-10}}{10^{-8}} \text{ F} = 2.66 \times 10^{-13} \text{ F}$$

(a) With a potential difference of 0.085 V, the net charge on either side of the wall is

$$Q = CV = 2.66 \times 10^{-13} \times 0.085 = 2.26 \times 10^{-4} \; C$$

(b) The number of K^+ ions required to produce this charge is

$$\frac{Q}{e} = \frac{2.26 \times 10^{-14}}{1.602 \times 10^{-19}} = 1.41 \times 10^{5}$$

The number of ions inside the cell is

$$0.155 \times 10^{-12} \times 6.022 \times 10^{23} = 9.33 \times 10^{10}$$

The fraction of ions at the surface is therefore

$$\frac{1.41 \times 10^{5}}{9.33 \times 10^{10}} = 1.5 \times 10^{-6}$$

16. At the left-hand electrode: $\frac{1}{2}H_2 \rightarrow H^+ + e^-$

At the right-hand electrode:

$$CrSO_4(s) + 2e^- \rightarrow SO_4^{2-} + Cr(s)$$

Overall reaction $CrSO_4(s) + H_2 \rightarrow 2H^+ + SO_4^{2-} + Cr(s)$

$$E^\circ = -0.40 \text{ V and } z = 2$$

(a) $E = E^\circ - \frac{RT}{2F} \ln ([H^+]^2[SO_4^{2-}])^u$

$$E/\text{V} = -0.40 - \frac{0.059}{2} \log_{10} (0.002)^2(0.001)$$

$$E = -0.152 \text{ V}$$

(b) The ionic strength is

$$I = \frac{1}{2}\left\{0.002 + (0.001 \times 4)\right\} = 0.003 \; M$$

$$\log_{10} y_{\pm} = -2 \times 0.51 \sqrt{0.003}$$
$$= -0.0559$$
$$y_{\pm} = 0.879$$
$$E = E^{\circ} - \frac{RT}{2F}\left\{\ln([H^+]^2[SO_4^{2-}])^u + \ln y_+^2 y_-\right\}$$
$$= E^{\circ} - \frac{RT}{2F}\left\{\ln([H^+]^2[SO_4^{2-}])^u + \ln y_{\pm}^3\right\}$$
$$E/V = -0.152 - \frac{0.059}{2}\log_{10}(0.879)^3$$

$$E = -0.152 + 0.05 = -0.147$$

17. At the left-hand electrode: $Cu(s) \rightarrow Cu^{2+} + 2e^-$

At the right-hand electrode:

$$AgCl(s) + e^- \rightarrow Ag(s) + Cl^-$$

Overall reaction: $2AgCl(s) + Cu(s) \rightarrow 2Ag(s) + Cu^{2+} + 2Cl^-$, $z = 2$

To a good approximation it can be assumed that the activity coefficients are unity at 10^{-4} M (DHLL gives $y_{\pm} = 0.988$). Then

$$E = E^{\circ} - \frac{RT}{zF}\ln([Cu^{2+}][Cl^-]^2)^u$$
$$0.191 = E^{\circ}/V - \frac{0.059}{2}\log_{10}\left\{10^{-4}(2 \times 10^{-4})^2\right\}^u$$
$$= E^{\circ}/V + 0.336$$
$$E^{\circ} = -0.145 \text{ V}$$

Suppose that at 0.20 M the activity coefficients are y_+ and y_-. Then

$$-0.074 = -0.145 - \frac{0.059}{2}\log_{10}\left\{0.20(0.40)^2\right\} - \frac{0.059}{2}\log_{10} y_+ y_-^2$$
$$-0.074 = -0.145 + 0.044 - 0.0295 \log_{10} y_{\pm}^3$$

$\log_{10} y_\pm = -0.294$

$y_\pm = 0.50$

18. (a) The anode reaction:

$2Tl(s) + 2Cl^-(0.02\ m) \rightarrow 2TlCl(s) + 2e^-$

The cathode reaction: $Cd^{2+}(0.01\ m) + 2e^- \rightarrow Cd(s)$

The overall **cell** reaction:

$2Tl(s) + Cd^{2+}(0.01\ m) + 2Cl^-(0.02\ m) \rightarrow 2TlCl(s) + Cd(s)$; $z = 2$

(b) This can alternatively be written as

(2) $2Tl(s) + Cd^{2+}(0.01\ m) \rightarrow 2Tl^+(\text{in } 0.01\ m\ CdCl_2) + Cd(s)$; $z = 2$

for which the Nernst equation is

$$E = E^\circ_{Cd^{2+}|Cd} - E^\circ_{Tl^+|Tl} - \frac{0.059\text{ V}}{2} \log([Tl^+]^2/[Cd^{2+}])^u$$

$E^\circ = -0.40 - (-0.34) = -0.06$ V

When $a_{CdCl_2} = 1$, E_1 is the value of E, and substituting $[Tl^+] = K_{sp}/[Cl^-]$ we have

$$E_1^\circ/\text{V} = -0.06 - 0.0295 \log(K_{sp}^2/[Cd^{2+}][Cl^-]^2)^u$$

$$= -0.06 - 0.0295 \log (1.6 \times 10^{-3})^2$$

$$E_1^\circ = -0.06 - (-0.165) = 0.105 \text{ V}$$

When $m = 0.01\ m$

$$E/\text{V} = -0.06 - 0.0295 \log (1.6 \times 10^{-3})^2/(0.01)(.02)^2$$

$$E = -0.06 - (-0.0057) = -0.054 \text{ V}$$

19. We are given that

(1) $\text{fumarate}^{2-} + 2H^+ + 2e^- \rightarrow \text{succinate}^{2-}$;

$$E^{\circ\prime} = 0.031 \text{ V}$$

(2) $\text{pyruvate}^- + 2H^+ + 2e^- \rightarrow \text{lactate}^-$

$$E^{\circ\prime} = -0.185 \text{ V}$$

Subtraction of (2) from (1) gives

$\text{fumarate}^{2-} + \text{lactate}^- \rightarrow \text{succinate}^{2-} + \text{pyruvate}^-$

$$E^{\circ\prime} = 0.216 \text{ V}; \; z = 2$$

(Note that this is also E°, the hydrogen ions having cancelled out). The Gibbs energy change is

$$\Delta G^\circ = -zFE^\circ = -2 \times 96\,487 \times 0.216 =$$

$$-41\,680 \text{ J mol}^{-1} = -41.7 \text{ kJ mol}^{-1}$$

The entropy change is obtained by use of Eq. 8.23:

$$\Delta S^\circ = 2 \times 96\,487 \times 2.18 \times 10^{-5} \text{ J K}^{-1} \text{ mol}^{-1}$$

$$= 4.21 \text{ J K}^{-1} \text{ mol}^{-1}$$

The enthalpy change can be calculated by use of Eq. 8.25, or more easily from the ΔG° and ΔS° values:

$$\Delta H^\circ = \Delta G^\circ + T\Delta S^\circ$$

$$= 41\,680 + (298.15 \times 4.207) \text{ J mol}^{-1}$$

$$= -40\,430 \text{ J mol}^{-1} = -40.4 \text{ kJ mol}^{-1}$$

20. The individual electrode processes are

$\text{Cd(Hg)} \rightarrow Cd^{2+} + 2e^-$

$$Hg_2^{2+} + 2e^- \rightarrow 2Hg$$

and the overall reaction is

$$Cd(Hg) + Hg_2^{2+} \rightarrow Cd^{2+} + 2Hg; \; z = 2$$

Since the solution is saturated with $Hg_2SO_4 \cdot \frac{8}{3}H_2O$ the overall reaction can be written as

$$Cd(Hg) + Hg_2SO_4(s) + \frac{8}{3}H_2O(\ell) \rightarrow CdSO_4 \cdot \frac{8}{3}H_2O(s) + 2Hg(\ell)$$

$$\Delta G^\circ = -2 \times 96\ 487 \times 1.01832 = -196\ 510 \text{ J mol}^{-1}$$

$$= -196.5 \text{ kJ mol}^{-1}$$

$$\Delta S^\circ = 2 \times 96\ 487 \times (-5.00 \times 10^{-5}) = -9.65 \text{ J K}^{-1} \text{ mol}^{-1}$$

$$\Delta H^\circ = -196\ 510 - (9.65 \times 298.15) = -199\ 390 \text{ J mol}^{-1}$$

$$= -199.4 \text{ kJ mol}^{-1}$$

21. $E^{\circ\prime} = E^\circ = 0.031 + 0.197 = 0.228$ V

$$\Delta G^\circ = -2 \times 96\ 487 \times 0.228 = 44\ 000 \text{ J mol}^{-1}$$

(a) If x kJ mol^{-1} = ΔG_f° (fumarate),

$$-690.44 - 139.08 - x + 181.75 = -44.00$$

$$x = -603.8$$

$$\Delta G_f^\circ = -603.8 \text{ kJ mol}^{-1}$$

(b) $\Delta S^\circ = 2 \times 96\ 487 \times (-1.45 \times 10^{-4})$

$$= -28.0 \text{ J K}^{-1} \text{ mol}^{-1}$$

$$\Delta H^\circ = \Delta G^\circ + T\Delta S^\circ = -44\ 000 - (298.15 \times 28.0)$$

$$= -52\ 350 \text{ J mol}^{-1}$$

If y kJ mol^{-1} = ΔH_f° (fumarate),

$$-908.68 - 210.66 - y + 287.02 = 52.35$$

$$y = -780.0$$

$\Delta H_f^\circ = -780.0 \text{ kJ mol}^{-1}$

22. (a) The cell reaction can be written

$Tl + H^+(a = 1) = Tl^+(\text{in HBr}; a = 1) + \frac{1}{2}H_2(1 \text{ atm})$

$E = E^\circ_{H^+/H_2} - E^\circ_{Tl^+/Tl} - 0.059 \log ([Tl^+]/[H^+])^u$

Since $Tl^+ = K_{sp}/[Br^-]$

$E/V = 0.34 - 0.059 \log (K_{sp}/[H^+][Br^-])^u$

$= 0.34 - 0.059 \log K_{sp}^u = 0.34 + 0.236$

$E = 0.58 \text{ V}$

(b) ΔH for the reaction is ΔH° for the half cell

$Tl \rightarrow Tl^+ + e^-$

$\Delta H^\circ = -zF(E - T dE/dT)$

$= -96\ 500\ [(0.34 - 298.15(-0.003)]$

$= -96\ 500\ (1.234) = -119 \text{ kJ mol}^{-1}$

23. Subtraction of the second reaction from the first gives

$AgBr(s) \rightarrow Ag^+ + Br^- \qquad E^\circ = -0.7278 \text{ V}; z = 1$

Then

$-0.7278 = 0.059\ 16 \log_{10}([Ag^+][Br^-])^u$

$\log_{10}([Ag^+][Br^-])^u = -12.30$

$K_{sp} = [Ag^+][Br^-] = 5.01 \times 10^{-13} \text{ mol}^2 \text{ kg}^{-2}$

Solubility $= \sqrt{K_{sp}} = 7.078 \times 10^{-7} \text{ mol kg}^{-1}$

24. The standard emf of the AgCl|Ag electrode is 0.2224 V and the cell reaction is $AgCl(s) + \frac{1}{2}H_2 \rightarrow H^+ + Cl^- + Ag(s); z = 1$

$$E = E^\circ - \frac{RT}{F}\ln(a_{H^+}a_{Cl^-})^u$$
$$\approx E^\circ - \frac{2RT}{F}\ln a^u_{H^+}$$

$$0.517 = 0.2224 - 2 \times 0.059\,16 \log_{10} a^u_{H^+}$$
$$\log_{10} a^u_{H^+} = \frac{0.2224 - 0.517}{2 \times 0.059\,16} = -2.49$$
pH = 2.49

25. At the two electrodes:

$$e^- + AgI(s) \rightarrow Ag(s) + I^-$$
$$Ag(s) \rightarrow Ag^+ + e^-$$

Overall reaction: $AgI(s) \rightarrow Ag^+ + I^-$; $z = 1$

$$E = \frac{RT}{F}\ln[m_{Ag^+} \times m_{Cl^-}]^u = \frac{RT}{F}\ln K_{sp}$$

$$\therefore -0.9509 = 0.059\,16 \log_{10} K^u_{sp}$$
$$K_{sp} = 8.45 \times 10^{-17}\ \text{mol}^2\ \text{kg}^{-2}$$
Solubility $= \sqrt{K_{sp}} = 9.19 \times 10^{-9}$ mol kg^{-1}

26. (a) The electrical work is $-\Delta G$;

$$\Delta G^\circ = \Delta H^\circ - T\Delta S^\circ$$
$$\Delta G^\circ/\text{J mol}^{-1} = -2\,877\,000 - 298.15 \times (-432.7)$$
$$= -2\,748\,000\ \text{J mol}^{-1} = -2750\ \text{kJ mol}^{-1}$$

Electrical work available = 2750 kJ mol^{-1}

(b) The total work done is $-\Delta A$

$$\Delta G^\circ = \Delta A^\circ - \Sigma\nu RT$$
$$\Sigma\nu = 4 - 1 - \frac{13}{2} = -3.5$$
$$\Sigma\nu RT = -3.5 \times 298.15 \times 8.314 = -8676\ \text{J mol}^{-1}$$
$$\Delta A^\circ = -2748 + 8.68 = -2739\ \text{kJ mol}^{-1}$$

Total available work = 2740 kJ mol^{-1}

27. The reactions at the two electrodes are

$\frac{1}{2}H_2$ (1 atm) → H^+ (0.1 *m*) + e^-

H^+ (0.2 *m*) + e^- → $\frac{1}{2}H_2$ (10 atm)

Every H^+ ion produced in the left-hand solution will have to pass through the membrane, to preserve electrical neutrality

H^+ (0.1 *m*) → H^+ (0.2 *m*)

The net reaction is $\frac{1}{2}H_2$ (1 atm) → $\frac{1}{2}H_2$ (10 atm)

$$E = \frac{RT}{F} \ln \frac{1}{\sqrt{10}} = (0.059\ 16\ \text{V}) \log_{10} \frac{1}{\sqrt{10}}$$

$$= -0.0296\ \text{V} = -2.96\ m\text{V}$$

28. The processes are:

$AgCl(s) + e^- \rightarrow Ag(s) + Cl^-$ (0.1 *m*)

$Ag(s) + Cl^-$(0.01 *m*) → $AgCl(s) + e^-$

The electrical neutrality is maintained by the passage of H^+ ions from right to left:

H^+(0.01 *m*) → H^+(0.10 *m*)

The overall process is

H^+(0.01 *m*) + Cl^-(0.01 *m*) → H^+(0.10 *m*) + Cl^-(0.10 *m*)

The emf is

$$E = -\frac{RT}{F} \ln \frac{0.1 \times 0.1}{0.01 \times 0.01}$$

$$E/\text{V} = -0.059 \log_{10} 100$$

$$E = -0.118\ \text{V}$$

29. (a) The processes at the electrodes are:

$\frac{1}{2}H_2 \rightarrow H^+(m_1) + e^-$

$H^+(m_2) + e^- \rightarrow \frac{1}{2}H_2$

To maintain electrical neutrality of the solutions, for every mole of H^+ produced in the left-hand solution, t_+ mol of H^+ ions will cross the membrane from left to right, and t_- mol of Cl^- ions will pass from right to left.

In the left-hand solution there is therefore a net gain of

$$(1 - t_+) \text{ mol} = t_- \text{ mol of } H^+$$

and of t_- mol of Cl^-

In the right-hand solution the net loss is

$$(1 - t_+) \text{ mol} = t_- \text{ mol of } H^+$$

and t_- mol of Cl^-.

The overall process is thus

$$t_-H^+(m_2) + t_-Cl^-(m_2) \rightarrow t_-H^+(m_1) + t_-Cl^-(m_1)$$

The emf is

$$E = -\frac{RT}{F} \ln \frac{m_1^{t_-} m_1^{t_-}}{m_2^{t_-} m_2^{t_-}}$$

$$= \frac{2t_-RT}{F} \ln \frac{m_2}{m_1}$$

(b) For $m_1 = 0.01\ m$ and $m_2 = 0.10\ m$,

$$0.0190 = 2t_- \times 0.059 \log_{10} 10$$

$$t_- = 0.161$$

$$t_+ = 0.839$$

30. (a) $M = M^+ (0.1\ m) + e^-$

$$\text{cell reaction:}\quad \frac{H^+(0.2\ m) + e^- = \frac{1}{2}H_2(1\ \text{atm})}{H^+(0.2\ m) + M = M^+(0.1\ m) + \frac{1}{2}H_2(1\ \text{atm})}$$

$$E^\circ = E^\circ_{H^+|H_2} - E^\circ_{M^+|M}$$

The Nernst equation is

$$-0.4 = E^\circ_{M^+|M}/V - 0.059 \log\frac{(0.1)}{(0.2)} = -E^\circ_{M^+|M}/V + 0.018$$

$$E^\circ_{M^+|M} = 0.418\ V$$

Upon addition of KCl almost all of the M^+ precipitates, and 0.10 m Cl^- is in excess. The value of M^+ in solution is found from the K_{sp}, namely,

$$K_{sp} = [M^+][Cl^-] = (M^+)(0.10\ M)$$

and from the Nernst equation

$$-0.1 = -E^\circ_{M^+|M} - 0.059 \log\frac{K_{sp}/(0.10)}{0.2}$$

$$-0.1 = -0.418 - 0.059 \log K^u_{sp} + 0.059 \log 0.02$$

$$-0.1 + 0.418 + 0.100 = -0.059 \log K^u_{sp}$$

$$\log K^u_{sp} = -\frac{0.418}{0.059} = -7.08;\ K_{sp} = 8.2 \times 10^{-8}\ \text{mol}^2\ \text{kg}^{-2}$$

CHAPTER 9: WORKED SOLUTIONS
KINETICS OF ELEMENTARY REACTIONS

1. (a) 3

   (b) Both rates are 3.6×10^{-3} mol dm^{-3} s^{-1}

   (c) None

   (d) Rate of Br^- disappearance decreased by a factor of 8; no effect on rate constant.

2. $\alpha = 1$, $\beta = 1$

$$k = \frac{5.0 \times 10^{-7} \text{ mol dm}^{-3} \text{ s}^{-1}}{(3.5 \times 10^{-2} \text{ mol dm}^{-3})(2.3 \times 10^{-2} \text{ mol dm}^{-3})}$$

$$= 6.21 \times 10^{-4} \text{ dm}^3 \text{ mol}^{-1} \text{ s}^{-1}$$

3. $\alpha = 2$, $\beta = 1$

$$k = \frac{7.4 \times 10^{-9} \text{ mol dm}^{-3} \text{ s}^{-1}}{(1.4 \times 10^{-2} \text{ mol dm}^{-3})^2 (2.3 \times 10^{-2} \text{ mol dm}^{-3})}$$

$$= 1.64 \times 10^{-3} \text{ dm}^6 \text{ mol}^{-2} \text{ s}^{-1}$$

4. (a) Half life, $t_{1/2} = \ln 2/k$

$$= \frac{0.6931}{3.72 \times 10^{-5} \text{ s}^{-1}} = 18\ 633 \text{ s}$$

$$= 5.18 \text{ hours}$$

   (b) Fraction undecomposed, $\dfrac{a_0 - x}{a_0} = e^{-kt}$

$$= e^{-3.72 \times 10^{-5} \times 3 \times 3600} = 0.669$$

5. $\dfrac{a_0 - x}{a_0} = e^{-kt}$

For half life, $t_{1/2}$: $0.5 = \exp(-kt_{1/2})$

For time $t_{99\%}$ to go to 99% completion:

$0.01 = \exp(-kt_{99\%})$

$t_{99\%} = \frac{1}{k} \ln 100$; $t_{1/2} = \frac{1}{k} \ln 2$

$$\frac{t_{99\%}}{t_{1/2}} = \frac{\ln 100}{\ln 2} = \frac{4.605}{0.693} = 6.64$$

6. For a second-order reaction, $t_{1/2} = 1/a_0 k$ (Table 9:1)

(a) $t_{1/2} = \dfrac{1}{1.3 \times 10^{11} \times 10^{-1}} = 7.7 \times 10^{-11}$ s = 77 ps

(b) $t_{1/2} = \dfrac{1}{1.3 \times 10^{11} \times 10^{-4}} = 7.7 \times 10^{-9}$ s = 7.7 ns

7. $\ln \dfrac{a_0}{a_0 - x} = -\ln$ (fraction remaining) $= kt$

$$t_{1/2} = \frac{\ln 2}{k}; \quad k = \frac{\ln 2}{t_{1/2}} = 0.0247 \text{ years}^{-1}$$

(a) After 25 years,

ln (fraction remaining) = −0.0247 x 25

fraction remaining = 0.539

quantity remaining = 0.539 μg

(c) After 70 years

ln (fraction remaining) = −0.0247 x 70 = −1.729

fraction remaining = 0.177

quantity remaining = 0.177 μg

8.

	$2NO$	+ Cl_2	→ $2NOCl$
Initial amounts:	5	2	0 mol
Amounts when half of the Cl_2 has reacted:	3	1	2 mol

The rate is then equal to

$(3/5)^2(1/2) \times 2.4 \times 10^{-3}$

$= 4.32 \times 10^{-4}$ mol dm^{-3} s^{-1}

9.

% decomposed	v/mmHg/min	$\log_{10}$(% remaining)	$\log_{10}\left(\frac{v/\text{mmHg}}{\text{min}}\right)$
0	8.53	2.00	0.931
5	7.49	1.98	0.874
10	6.74	1.95	0.827
15	5.90	1.93	0.771
20	5.14	1.90	0.711
25	4.69	1.88	0.671
30	4.31	1.85	0.634
35	3.75	1.81	0.574
40	3.11	1.78	0.493
45	2.67	1.74	0.427
50	2.29	1.70	0.360

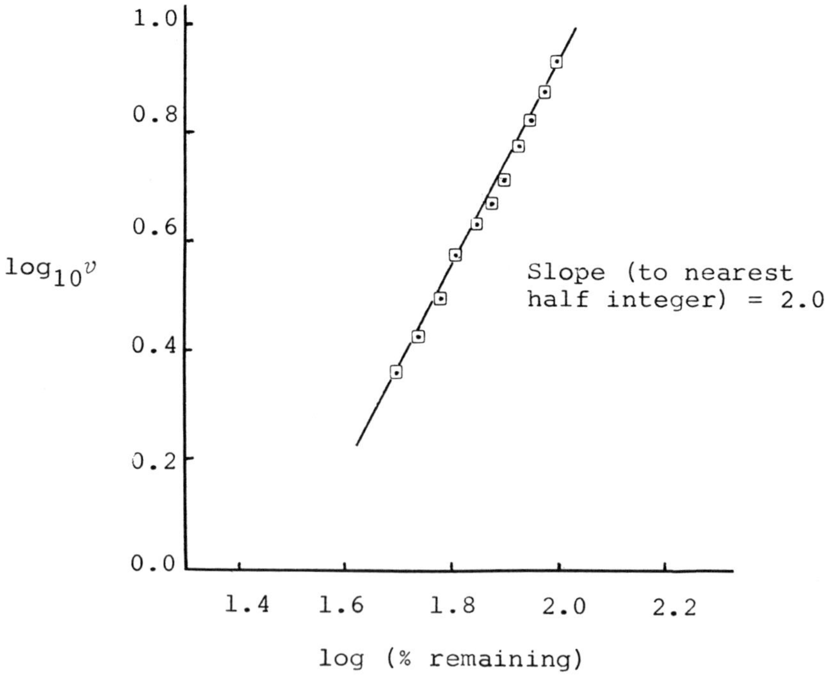

Slope of plot of $\log_{10}(v/\text{mmHg min}^{-1})$ against $\log_{10}$ (% remaining) (to nearest half integer) = 2.0. This is the order with respect to time.

10. Half life $t_{1/2} = 14.3 \times 24 \times 60 \times 60$ s $= 1.236 \times 10^6$ s

$$k = \frac{0.693}{t_{1/2}} = 5.61 \times 10^{-7}\ \text{s}^{-1}$$

$$= \frac{0.693}{14.3}\ \text{days}^{-1} = 0.0485\ \text{days}^{-1}$$

(a) After 10 days

$$\ln \frac{n_0}{n} = 0.0485 \times 10 = 0.485; \quad \frac{n_0}{n} = 1.624$$

$$\text{Activity remaining} = \frac{n}{n_0} = 0.616 = 61.6\%$$

(b) After 20 days

$$\ln \frac{n_0}{n} = 0.0485 \times 20 = 0.97; \quad \frac{n_0}{n} = 2.64$$

$$\text{Activity remaining} = \frac{n}{n_0} = 0.379 = 37.9\%$$

(c) After 100 days

$$\ln \frac{n_0}{n} = 0.0485 \times 100 = 4.85; \quad \frac{n_0}{n} = 127.7$$

$$\text{Activity remaining} = \frac{n}{n_0} = 0.0078 = 0.78\%$$

11.

t/days	n	$\ln(n_0/n)$
0	4280 (n_0)	-
1	4245	8.21×10^{-3}
2	4212	16.0×10^{-3}
3	4179	23.9×10^{-3}
4	4146	31.8×10^{-3}
5	4113	39.8×10^{-3}
10	3952	79.7×10^{-3}
15	3798	119.5×10^{-3}

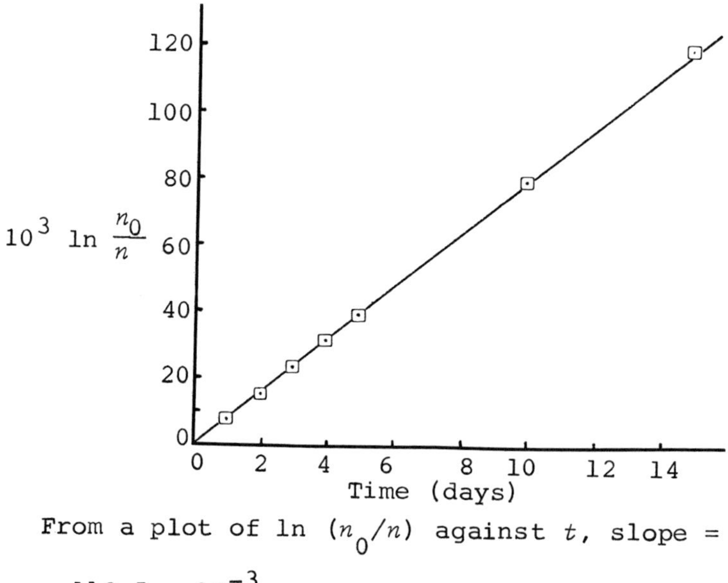

From a plot of $\ln(n_0/n)$ against t, slope = k

$$= \frac{119.5 \times 10^{-3}}{15} = 7.97 \times 10^{3} \text{ days}^{-1}$$

$$= \frac{7.97 \times 10^{-3}}{3600 \times 24} = 9.22 \times 10^{-8} \text{ s}^{-1}$$

$$t_{1/2} = \frac{0.693}{7.97 \times 10^{-3}} = 87.0 \text{ days}$$

(a) After 60 days, $\ln n_0/n = 0.478$

$$n_0/n = 1.613; \quad n = 2653$$

(b) After 365 days, $\ln n_0/n = 2.909$

$$n_0/n = 18.34; \quad n = 233$$

12. The rate equation is

$$k_1 t = \frac{x_e}{a_0} \ln \frac{x_e}{x_e - x} \quad \text{(Eq. 9.48)}$$

The time for half the equilibrium amount of product to be formed is given by putting $x = x_e/2$:

$$k_1 t_{1/2} = \frac{x_e}{a_0} \ln \frac{x_e}{x_e/2} = \frac{x_e}{a_0} \ln 2$$

The equilibrium constant is $x_e/(a_0 - x_e)$:

$$\frac{a_0 - x_e}{x_e} = \frac{1}{0.16}$$

$$\frac{a_0}{x_e} = \frac{1}{0.16} + 1 = 7.25$$

Thus $t_{1/2} = \dfrac{1}{7.25 \times 3.3 \times 10^{-4}\ \text{s}^{-1}} \ln 2 = 290\ \text{s}$

13. $T_1 = 293.15\ \text{K} \qquad 1/T_1 = 3.4112 \times 10^{-3}\ \text{K}^{-1}$

$T_2 = 303.15\ \text{K} \qquad 1/T_2 = 3.2987 \times 10^{-3}\ \text{K}^{-1}$

Slope of a plot of $\ln k$ against $1/T$ is

$$\frac{\ln 2}{(3.2987 - 3.4112) \times 10^{-3}\text{K}^{-1}} = -\frac{0.6931\ \text{K}}{0.1125 \times 10^{-3}}$$

$$= -6161\ \text{K}$$

$E = -8.314\ \text{J K}^{-1}\ \text{mol}^{-1} \times \text{slope} = 51\ 200\ \text{J mol}^{-1}$

$= 51.2\ \text{kJ mol}^{-1}$

14. $T_1 = 493.15$ K $\qquad 1/T_1 = 2.0278 \times 10^{-3}$ K^{-1}

$T_2 = 503.15$ K $\qquad 1/T_2 = 1.9875 \times 10^{-3}$ K^{-1}

Slope of a plot of ln k against $1/T$ is

$$-\frac{0.6931 \text{ K}}{0.0403 \times 10^{-3}} = -17\ 200 \text{ K}$$

$E = 143\ 000$ J mol^{-1} = 143.0 kJ mol^{-1}

15. Rate constant ratio = $e^{20\ 000/(8.314 \times T/\text{K})}$

(a) If T = 273.15 K, ratio = $e^{20\ 000/8.314 \times 273.15}$

$= 6.68 \times 10^3$

(b) If T = 1273.15 K, ratio = $e^{20\ 000/8.314 \times 1273.15}$

= 6.62

16. T_1 = 15°C = 288.15 K; $1/T_1 = 3.4704 \times 10^{-3}$ K^{-1}

T_2 = 25°C = 298.15 K; $1/T_2 = 3.3540 \times 10^{-3}$ K^{-1}

Slope of a plot of ln k against $1/T$ is

$$-\frac{0.6931}{0.1164 \times 10^{-3}} = 5.954 \times 10^3 \text{ K}$$

$E = 8.314 \times 5.954 \times 10^3$ J mol^{-1} = 49 500 J mol^{-1}

= 49.5 kJ mol^{-1}

17. ln 40 = 3.689

T_1 = 4°C = 277.15 K; $\qquad 1/T_1 = 3.6082 \times 10^{-3}$ K^{-1}

T_2 = 25°C = 298.15 K; $\qquad 1/T_2 = 3.3540 \times 10^{-3}$ K^{-1}

Slope of plot of ln k against $1/T$ $= -\dfrac{3.689}{0.2542 \times 10^{-3}}$

= −14 500 K

$E = 120\ 700$ J mol^{-1} = 120.7 kJ mol^{-1}

18.

T/K	$\dfrac{k \times 10^{-9}}{cm^6\ mol^{-2}s^{-1}}$	$\log_{10}\left(\frac{T}{K}\right)$	$\log_{10}\left(k \Big/ \frac{10^9\ cm^6}{mol^2\ s}\right)$
80.0	41.8	1.903	1.62
143.0	20.2	2.155	1.31
228.0	10.1	2.358	1.00
300.0	7.1	2.477	0.85
413.0	4.0	2.616	0.60
564.0	2.8	2.751	0.45

Slope of plot of $\log_{10}(k/10^9\ cm^6\ mol^{-2}\ s^{-1})$ against $\log_{10}(T/K)$, to the nearest half integer, is -1.5.

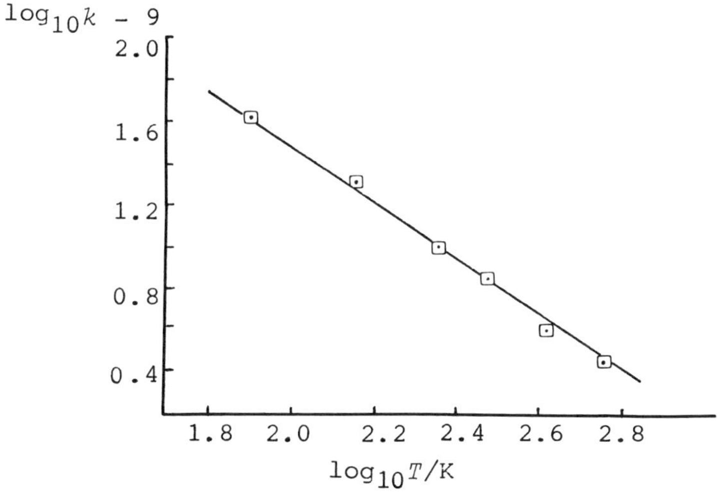

19. Rate constant ratio = $e^{50/8.314}$ = 409

20. $2.34 \times 10^{-2}\ dm^3 mol^{-1}s^{-1} = A\ e^{-150\,000/8.314 \times 673.15}$

$= A \times 2.291 \times 10^{-12}$

$A = 1.02 \times 10^{10}\ dm^3 mol^{-1}s^{-1}$

For a second-order reaction

$$\Delta H^{\ddagger} = E_{exp} - 2RT \qquad \text{(Eq. 9.100)}$$

$$= 150\ 000 - 2 \times 8.314 \times 673.15 \text{ J mol}^{-1}$$

$$= 138\ 806 \text{ J mol}^{-1} = 138.8 \text{ kJ mol}^{-1}$$

$$A = e^2 \frac{\mathbf{k}T}{h} e^{\Delta S^{\ddagger}/R} \qquad \text{(Eq. 9.101)}$$

$$1.02 \times 10^{10} = \frac{2.718^2 \times 1.381 \times 10^{-23} \times 673.15}{6.626 \times 10^{-34}} e^{\Delta S^{\ddagger}/R}$$

$$= 1.0365 \times 10^{14}\ e^{\Delta S^{\ddagger}/R}$$

$$9.84 \times 10^{-5} = e^{\Delta S^{\ddagger}/8.314 \text{ J K}^{-1} \text{ mol}^{-1}}$$

$$\frac{\Delta S^{\ddagger}}{8.314 \text{ J K}^{-1} \text{ mol}^{-1}} = -9.226$$

$$\Delta S^{\ddagger} = -76.71 \text{ J K}^{-1} \text{ mol}^{-1}$$

$$\Delta G^{\ddagger} = \Delta H^{\ddagger} - T\Delta S^{\ddagger} = 138\ 806 + 76.71 \times 673.15$$

$$= 190\ 400 \text{ J mol}^{-1} = 190.4 \text{ kJ mol}^{-1}$$

21. From the data:

T/°C	$10^3/(T/\text{K})$	$\log_{10}(k/\text{s}^{-1})$
15.2	3.468	-5.380
20.0	3.4112	-5.118
25.0	3.3540	-4.863
30.0	3.2987	-4.618
37.0	3.2243	-4.288

A graph of $\log_{10}k$ against $10^3/T$, gives a slope equal to -4.393×10^3K

$$E = 84\ 082 \text{ J mol}^{-1} = 84.1 \text{ kJ mol}^{-1}$$

$$\Delta H^{\ddagger} = 84\ 082 - 2479 = 81\ 603 \text{ J mol}^{-1}$$

$= 81.6\ \text{kJ mol}^{-1}$

$$\log_{10} k = \log_{10} \frac{\mathbf{k}T}{h} - \frac{\Delta G^{\ddagger}}{19.14T}$$

At 25.0°C, $\mathbf{k}T/h = 6.21 \times 10^{12}\ \text{s}^{-1}$, and $\log_{10}(k/\text{s}^{-1})$

$= -4.86$;

$$\text{Then } -4.86 = 12.79 - \frac{\Delta G^{\ddagger}}{5706.6\ \text{J mol}^{-1}}$$

and therefore

$\Delta G^{\ddagger} = 100\ 721\ \text{J mol}^{-1} = 100.7\ \text{kJ mol}^{-1}$

The pre-exponential factor can be calculated from

$$\log_{10} k = \log_{10} A - \frac{E}{(19.14\ \text{J K}^{-1}\ \text{mol}^{-1})T}$$

At 298.15 K,

$$-4.86 - \log_{10}(A/\text{s}^{-1}) - \frac{84\ 082}{5706.6}$$

$\log_{10}(A/\text{s}^{-1}) = 9.87$

$A = 7.48 \times 10^{9}\ \text{s}^{-1}$

$$\Delta S^{\ddagger} = \frac{81\ 603 - 100\ 721}{298.15} = -64.1\ \text{J K}^{-1}\ \text{mol}^{-1}$$

22.

T/K	k/s^{-1}	$10^3(T/\text{K})^{-1}$	$\log_{10}(k/\text{s}^{-1})$
313.05	4.67×10^{-6}	3.194	−5.331
316.95	7.22×10^{-6}	3.155	−5.141
320.25	10.0×10^{-6}	3.123	−5.000
323.35	13.9×10^{-6}	3.093	−4.857

Slope of plot of ln k against $1/T$ is -4.70×10^{3} K

$E = 89\ 958\ \text{J mol}^{-1} = 90.0\ \text{kJ mol}^{-1}$

$\Delta H^{\ddagger} = 89\ 958 - 2479 = 87\ 479\ \text{J mol}^{-1}$

$= 87.5\ \text{kJ mol}^{-1}$

$$\log_{10} k = \log_{10} \frac{\mathbf{k}T}{h} - \frac{\Delta G^{\ddagger}}{(19.14\ \text{J K}^{-1}\ \text{mol}^{-1})T}$$

At $40.0°\text{C} = 313.15\ \text{K},\ \frac{\mathbf{k}T}{h} = 6.527 \times 10^{12}\ \text{s}^{-1}$;

$\log_{10}\left\{\left(\frac{\mathbf{k}T}{h}\right)/\text{s}^{-1}\right\} = 12.81$

From graph, $\log(k/\text{s}^{-1})$ at $40.0° = -5.335$

$$-5.335 = 12.81 - \frac{\Delta G^{\ddagger}/\text{J mol}^{-1}}{19.14 \times 313.15}$$

$\Delta G^{\ddagger} = 108\ 783\ \text{J mol}^{-1} = 108.8\ \text{kJ mol}^{-1}$

$$\Delta S^{\ddagger} = \frac{87\ 479 - 108\ 783}{313.15} = -68.03\ \text{J K}^{-1}\ \text{mol}^{-1}$$

$$A = \text{e}\ \frac{\mathbf{k}T}{h}\ \text{e}^{\Delta S^{\ddagger}/R}$$

$$\log_{10}(A/\text{s}^{-1}) = \log_{10}\text{e} + 12.81 - \frac{68.03}{19.14}$$

$= 0.434 + 12.81 - 3.55 = 9.67$

$A = 4.94 \times 10^{9}\ \text{s}^{-1}$

23. $\ln \frac{3460}{530} = 1.876$

$T_1 = 333.15\ \text{K} \qquad 1/T_1 = 3.001\ 65 \times 10^{-3}\ \text{K}^{-1}$

$T_2 = 338.15\ \text{K} \qquad 1/T_2 = 2.957\ 27 \times 10^{-3}\ \text{K}^{-1}$

Slope of plot of $\ln(k_1/k_2)$ against $1/T$

$$= -\frac{1.876}{4.438 \times 10^{-5}\text{K}^{-1}} = -42\ 270\ \text{K}$$

$E = 351\ 403\ \text{J mol}^{-1} = 351.4\ \text{kJ mol}^{-1}$

$\Delta H^{\ddagger} = 351\ 403 - 2770 = 348\ 633\ \text{J mol}^{-1}$

$= 348.6\ \text{kJ mol}^{-1}$

At 60.0°C, k = (ln 2)/3460 s = 2.003×10^{-4} s^{-1}

$$\frac{\mathbf{k}T}{h} = 6.94 \times 10^{12}\ s^{-1}$$

$$2.003 \times 10^{-4} = 6.94 \times 10^{12} e^{-\Delta G^{\ddagger}/8.314 \times 333.15\ J\ mol^{-1}}$$

$$\Delta G^{\ddagger} = 105\ 470\ J\ mol^{-1} = 105.5\ kJ\ mol^{-1}$$

$$\Delta S^{\ddagger} = \frac{\Delta H^{\ddagger} - \Delta G^{\ddagger}}{T} = \frac{348\ 630 - 105\ 500}{333.15} J\ K^{-1}\ mol^{-1}$$

$$= 730\ J\ K^{-1}\ mol^{-1}$$

24. (a) Mass of HI molecule = $\dfrac{127.91 \times 10^{-3}\ kg\ mol^{-1}}{6.022 \times 10^{23}\ mol^{-1}}$

$$= 2.124 \times 10^{-25}\ kg$$

Then, from Eq. 9.73,

$$Z_{AA} = 2 \times (0.35 \times 10^{-9}\ m)^2 \times (6.022 \times 10^{23}\ m^{-3})^2 \times \left(\frac{\pi \times 1.381 \times 10^{-23}\ J\ K^{-1} \times 573.15\ K}{2.124 \times 10^{-25}\ kg}\right)^{1/2}$$

$$= 3.040 \times 10^{31}\ m^{-3}\ s^{-1}$$

(b) Rate of reaction under these conditions with $E = 184\ kJ\ mol^{-1}$, is

$$3.040 \times 10^{31} \exp(-184\ 000/8.314 \times 573.15) m^{-3}\ s^{-1}$$

$$= 5.167 \times 10^{14}\ m^{-3}\ s^{-1} = 8.58 \times 10^{-10}\ mol\ m^{-3}\ s^{-1}$$

The rate constant is this rate divided by the square of the concentration, which is 1 mol m^{-3};

$$k = \frac{8.58 \times 10^{-10}\ mol\ m^{-3}\ s^{-1}}{(1\ mol\ m^{-3})^2}$$

$$= 8.58 \times 10^{-10}\ m^3\ mol^{-1}\ s^{-1}$$

$$= 8.58 \times 10^{-7}\ dm^3\ mol^{-1}\ s^{-1}$$

The collision frequency factor (the pre-exponential factor) is k divided by exp (-184 000/8.314 x 573.15), and is

$$5.048 \times 10^{10} \text{ dm}^3 \text{ mol}^{-1} \text{ s}^{-1}$$

To obtain the entropy of activation we use the relationship

$$A = e^2 \frac{\mathbf{k}T}{h} e^{\Delta S^\ddagger/R} \qquad \text{(Eq. 9.101)}$$

With $A = 5.048 \times 10^{10}$ dm^3 mol^{-1} s^{-1} this gives

$$5.048 \times 10^{10} = e^2 \frac{\mathbf{k}T}{h} e^{\Delta S^\ddagger/R}$$

$$= \frac{2.718^2 \times 1.381 \times 10^{-23} \times 573.15}{6.626 \times 10^{-34}} e^{\Delta S^\ddagger/R}$$

$$e^{\Delta S^\ddagger/R} = 5.719 \times 10^{-4}$$

$$\frac{\Delta S^\ddagger}{R} = -7.47$$

$$\Delta S^\ddagger = -62.1 \text{ J K}^{-1} \text{ mol}^{-1}$$ (standard state: 1 mol dm^{-3})

25.

$\dfrac{I}{10^{-3} \text{ mol dm}^{-3}}$	$\dfrac{k}{\text{dm}^3 \text{ mol}^{-1} \text{ s}^{-1}}$	$\sqrt{\dfrac{I}{\text{mol dm}^{-1}}}$	$\log_{10}\left(\dfrac{k}{\text{dm}^3 \text{ mol}^{-1} \text{ s}^{-1}}\right)$
2.45	1.05	0.0495	0.0212
3.65	1.12	0.0604	0.0492
4.45	1.16	0.0667	0.0645
6.45	1.18	0.0815	0.0719
8.45	1.26	0.0941	0.1004
12.45	1.39	0.1116	0.1430

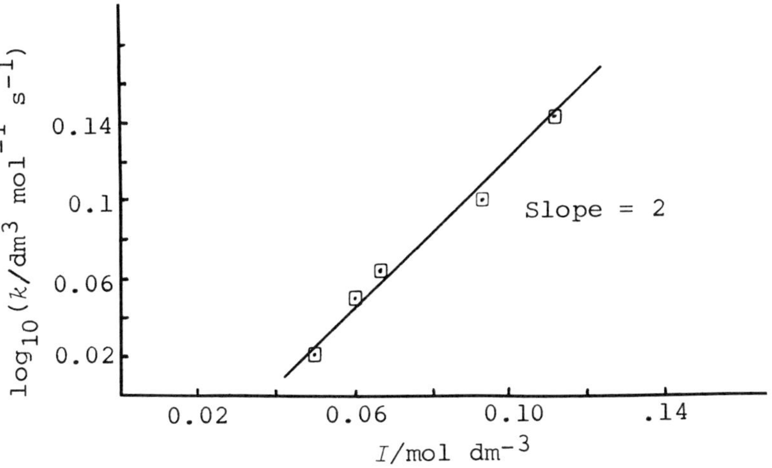

Slope of plot of $\log_{10}(k/\mathrm{dm^3\ mol^{-1}\ s^{-1}})$ against $(\sqrt{I/\mathrm{mol\ dm^{-3}}})$ is approximately 2:

$$z_A z_B = 2$$

26. Ionic strengths of the five mixtures are:

1) $\frac{1}{2}[4 \times 5.0 \times 10^{-4} + 1.0 \times 10^{-3} + (2 \times 7.95 \times 10^{-4})]$
$= 2.295 \times 10^{-3}\ M$

2) $\frac{1}{2}[4 \times 5.96 \times 10^{-4} + 11.92 \times 10^{-4} +$
$(2 \times 1.004 \times 10^{-3})] = 2.79 \times 10^{-3}\ M$

3) $\frac{1}{2}[4 \times 6.0 \times 10^{-4} + 12.0 \times 10^{-4} +$
$(2 \times 0.696 \times 10^{-3}) + 0.01] = 7.496 \times 10^{-3}\ M$

4) $\frac{1}{2}(24.0 \times 10^{-4} + 12.0 \times 10^{-4} + 1.392 \times 10^{-3} +$
$0.04) = 22.50 \times 10^{-3}\ M$

5) $\frac{1}{2}(24.0 \times 10^{-4} + 12.0 \times 10^{-4} + 1.392 \times 10^{-3} +$
$0.06) = 32.5 \times 10^{-3}\ M$

$\frac{I}{10^{-3}\ \text{mol cm}^{-3}}$	$\frac{k}{\text{dm}^3\ \text{mol}^{-1}\ \text{s}^{-1}}$	$\sqrt{\frac{I}{\text{mol dm}^{-3}}}$	$\log_{10}\frac{k}{\text{dm}^3\ \text{mol}^{-1}\ \text{s}^{-1}}$
2.295	1.52	0.0479	0.1818
2.790	1.45	0.0528	0.1614
7.496	1.26	0.0866	0.0899
22.50	0.97	0.1500	-0.0132
32.5	0.91	0.1803	-0.0409

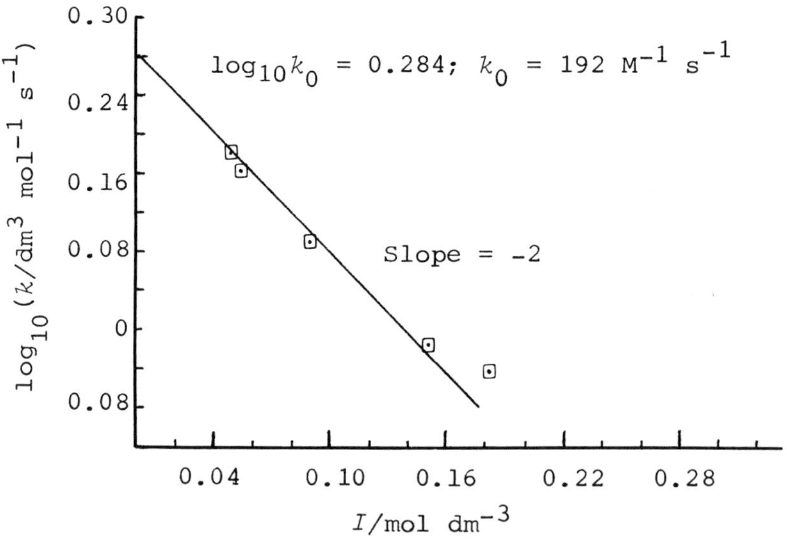

Plot of $\log_{10}(k/\text{dm}^3\ \text{mol}^{-1}\ \text{s}^{-1})$ against $\sqrt{I/(\text{mol dm}^{-3})}$ extrapolates to

$$\log_{10}(k_0/\text{dm}^3\ \text{mol}^{-1}\ \text{s}^{-1}) = 0.284$$

$$k_0 = 1.92\ \text{dm}^3\ \text{mol}^{-1}\ \text{s}^{-1}$$

The slope is -2, which is consistent with $z_A z_B = -2$

27. $K^{\ddagger} = \frac{[X^{\ddagger}]}{[A][B]} \frac{y^{\ddagger}}{y_A y_B}$

Suppose that $v = k\,[X^{\ddagger}]\,y^{\ddagger}$; then

$v = K^{\ddagger}\,k\,[A][B]\,y_A y_B$

and

$k = K^{\ddagger}\,k\,y_A y_B = k_0 y_A y_B$

$\log_{10} k = \log_{10} k_0 + \log_{10} y_A y_B$

$= \log_{10} k_0 - B[z_A{}^2 + z_B{}^2]\sqrt{I}$

Plots of $\log_{10} k$ against $\sqrt{I}$ will always have negative slopes; this conclusion is inconsistent with the results in Fig. 9.18.

28. $\ln \frac{k}{k_0} = \frac{\Delta V^{\ddagger}}{RT} P$ (Eq. 9.128)

1999 atm = 2.025×10^8 Pa

$\ln 2 = -\frac{(\Delta V^{\ddagger}/\text{m}^3) \times 2.025 \times 10^8}{8.314 \times 300 \quad \text{mol}^{-1}}$

$= -81\,188 \times (\Delta V^{\ddagger}/\text{m}^3\ \text{mol}^{-1})$

$\Delta V^{\ddagger} = -8.538 \times 10^{-6}\ \text{m}^3\ \text{mol}^{-1} = -8.54\ \text{cm}^3\ \text{mol}^{-1}$

29.

Pressure/10^2 kPa	1.00	345	689	1033
$k/10^{-6}\ \text{s}^{-1}$	7.18	9.58	12.2	15.8
$\ln(k/\text{s}^{-1})$	-11.84	-11.56	-11.31	-11.06

From plot of ln k against pressure, slope $= 7.55 \times 10^{-9}\ \text{Pa}^{-1}$

$7.55 \times 10^{-9} = \frac{\Delta V^{\ddagger}/\text{m}^3\ \text{mol}^{-1}}{8.314 \times 298.15}$

$\Delta V^{\ddagger} = -1.87 \times 10^{-5}\ \text{m}^3\ \text{mol}^{-1}$

$= -18.7 \text{ cm}^3 \text{ mol}^{-1}$

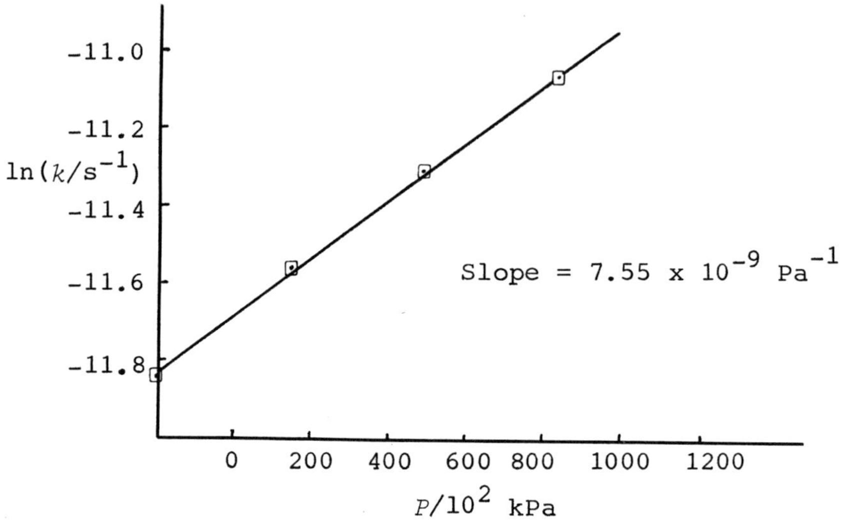

30.

P/kPa	$k/10^4$ dm^3 mol^{-1} s^{-1}	ln($k/10^4$ dm^3 mol^{-1} s^{-1})
101.3	9.30	2.23
2.76×10^4	11.13	2.41
5.51×10^4	13.1	2.57
8.27×10^4	15.3	2.73
11.02×10^4	17.9	2.88

Slope of plot of ln k against pressure = 5.75×10^{-9} Pa^{-1}

$$5.75 \times 10^{-9} = -\frac{\Delta V^{\ddagger}/\text{m}^3 \text{ mol}^{-1}}{8.314 \times 298.15}$$

$$\Delta V^{\ddagger} = -1.43 \times 10^{-5} \text{ m}^3 \text{ mol}^{-1}$$

$$= -14.3 \text{ cm}^3 \text{ mol}^{-1}$$

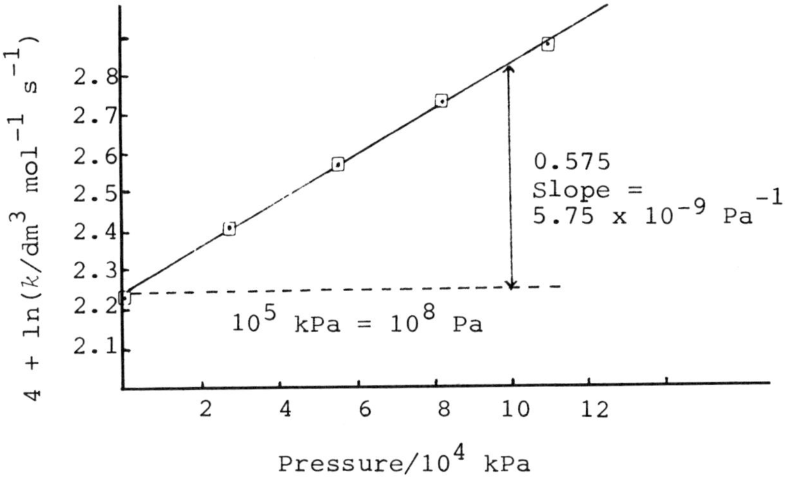

31. $\frac{dx}{dt} = k(a_0 - x)^n$

which integrates to

$$k = \frac{1}{t(n - 1)} \frac{1}{(a_0 - x)^{(n-1)}} - \frac{1}{a_0{}^{(n-1)}}$$

With $x = a_0/2$,

$$t_{1/2} = \frac{1}{k(n - 1)}\left[\frac{1}{(a_0/2)^{(n-1)}} - \frac{1}{a_0{}^{(n-1)}}\right]$$

$$= \frac{2^{(n-1)} - 1}{k\ a_0{}^{(n-1)}\ (n - 1)}$$

32. When the concentration has reached nc, where c is the concentration produced by one dose, the concentration will fall to $(n - 1)c$ during the interval between successive doses. The next dose restores the concentration to $(n - 1)c + c = nc$. The "steady state"

has therefore been reached.

33.

	2A	+	B	→	Z
Initially	a_0		b_0		0
After time t	$a_0 - 2x$		$b_0 - x$		x

$$\frac{dx}{dt} = k(a_0 - 2x)(b_0 - x)$$

$$kdt = \frac{dx}{(a_0 - 2x)(b_0 - x)}$$

$$= \left\{\frac{2}{a_0 - 2x} - \frac{1}{(b_0 - x)}\right\} \frac{dx}{2b_0 - a_0}$$

$$kt = \frac{1}{2b_0 - a_0}\left\{- \ln (a_0 - 2x) + \ln (b_0 - x)\right\} + I$$

When $t = 0$, $x = 0$, and therefore

$$I = \frac{1}{2b_0 - a_0}\left\{ \ln a_0 - \ln b_0\right\}$$

$$kt = \frac{1}{2b_0 - a_0}\left\{ \ln \frac{a_0}{a_0 - 2x} - \ln \frac{b_0}{(b_0 - x)}\right\}$$

$$= \frac{1}{(2b_0 - a_0)}\left\{\ln \frac{a_0(b_0 - x)}{b_0(a_0 - 2x)}\right\}$$

34.

	2A	+	B	→	Z
Initially	$2a_0$		a_0		0
At time t	$2a_0 - 2x$		$a_0 - x$		x

$$\frac{dx}{dt} = k(2a_0 - 2x)^2(a_0 - x)$$

$$= 4k(a_0 - x)^3$$

$$\frac{dx}{(a_0 - x)^3} = 4k\ dt$$

$$\frac{1}{2(a_0 - x)^2} = 4kt + I$$

$x = 0$ when $t = 0$; $I = 1/(2a_0^2)$

$$\frac{1}{2(a_0 - x)^2} - \frac{1}{2a_0^2} = 4kt$$

$$\frac{2a_0x - x^2}{a_0^2(a_0 - x)^2} = 8kt$$

The half life $t_{1/2}$ is when $x = a_0/2$ (note that at this time half of A and half of B have ben consumed; if all reactants were not present in stoichiometric proportions the half lives would be different for the two reactants, and we could not speak of the half life of the reaction).

$$\frac{2a_0^2/2 - a_0^2/4}{a_0^2(a_0/2)^2} = 8kt_{1/2}$$

which reduces to

$$t_{1/2} = \frac{3}{8a_0^2\ k}$$

35. $\frac{d[Y]}{dt} = k_1\ [A]$; $\frac{d[Z]}{dt} = k_2\ [A]$

$$\frac{d[Y]}{d[Z]} = \frac{k1}{k_2}$$

Integrating

$$[Y] = \frac{k_1}{k_2}[Z] + I$$

At $t = 0$, $[Y] = [Z]$ and therefore $I = 0$. Thus

$$\frac{[Y]}{[Z]} = \frac{k_1}{k_2}$$

36. The rate of consumption of A is given by

$$-\frac{d[A]}{dt} = k_1 [A] \qquad (1)$$

The rates of formation of B and C are

$$\frac{d[B]}{dt} = k_1 [A] - k_2 [B] \qquad (2)$$

$$\frac{d[C]}{dt} = k_2 [B] \qquad (3)$$

The first equation may be integrated at once to give

$$[A] = [A]_0 \, e^{-kt} \qquad (4)$$

Insertion of this in Eq. (2) gives

$$\frac{d[B]}{dt} = k_1[A]_0 \, e^{-k_1 t} - k_2[B] \qquad (5)$$

With the boundary condition $t = 0$, $[B] = 0$

This integrates to

$$[B] = [A]_0 \frac{k_1}{k_2 - k_1}(e^{-k_1 t} - e^{-k_2 t}) \qquad (6)$$

$$[A]_0 = [A] + [B] + [C] \qquad (7)$$

and thus

$$[C] = [A]_0 - [A] - [B] \tag{8}$$

and insertion of Eqs. (4) and (6) leads to

$$[C] = [A]_0 \left\{ 1 + \frac{k_2\ e^{-k_1 t} - k_1\ e^{-k_2 t}}{k_1 - k_2} \right\}$$

37. (a) The rate equation, with the catalyst concentration incorporated in the rate constants, is

$$\frac{dx}{dt} = k_1(a_0 - x) - k_1 x^2$$

At equilibrium

$$k_1(a_0 - x_e) - k_1 x_e^2 = 0$$

where x_e is the concentration of Y and Z at equilibrium. Hence

$$k_{-1} = \frac{k_1(a_0 - x_e)}{x_e^2}$$

Insertion of this in the first equation leads, after rearrangement, to

$$\frac{x_e^2\ dx}{(a_0 - x)x_e^2 - (a_0 - x_e)x^2} = k_1\ dt$$

Integration of the left-hand side of this equation can be carried out after resolution into partial fractions:

$$\text{L.H.S.} = \frac{p}{x_0 - x} + \frac{q}{a_0 x_e + a_0 x - x_e x}$$

and it is found that

$$p = \frac{x_e}{2a_0 - x_e} \qquad q = \frac{x_e(a_0 - x_e)}{2a_0 - x_e}$$

The integration is then straight forward (but fairly lengthy); with the boundary condition $t = 0$, $x = 0$ the result is

$$k_1 = \frac{x_e}{(2a_0 - x_e)t} \ln \frac{a_0 x_e + x(a_0 - x_e)}{a_0(x_e - x)}$$

Readers wishing further mathematical details are referred to C. Capellos and B. H. T. Bielski, *Kinetic Systems* (New York: Wiley, Interscience, 1972) p. 41-43.

(b) The equilibrium constant $K = k_1/k_{-1}$ is given by

$$K = \frac{a_0 - x_e}{x_e^2}$$

and this quadratic can be solved to give x_e in terms of K. This expression can then be substituted into the equation for k_1.

(c) To deal with the numerical data, which are in terms of percent hydrolysis, it is convenient to define

$$r \equiv \frac{x}{a_0} \quad \text{and} \quad r_e = \frac{x_e}{a_0}$$

The integrated equation then becomes

$$k_1 = \frac{x_e}{(2 - r_e)t} \ln \frac{r_e + r(1 - r_e)}{r_e - r}$$

From the data we then have $r_e = 0.90$ and

t/s	r	$r_e + r(1 - r_e)$	$r_e - r$	$\ln \frac{r_e + r(1 - r_e)}{r_e - r}$
1350	0.212	0.9212	0.688	0.292
2070	0.307	0.9307	0.593	0.451
3060	0.434	0.9434	0.466	0.705
5340	0.595	0.9595	0.315	1.114
7740	0.7345	0.97345	0.1655	1.77

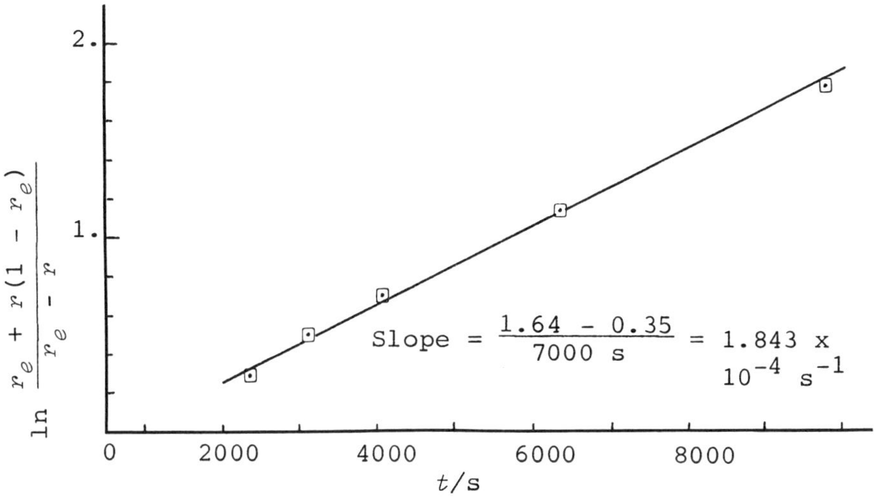

Slope of a plot of $\ln \frac{r_e + r(1 - r_e)}{r_e - r}$

against t is 1.843×10^{-4} s^{-1}

This slope is

$$\frac{k_1(2 - r_e)}{r_e} = 1.222\ k_1$$

Thus $k_1 = 1.51 \times 10^{-4}$ s^{-1}

This is the pseudo first-order rate constant for the reaction in the presence of 0.05 *M* HCl. The second-order rate constant is

$$\frac{1.51 \times 10^{-4}\ s^{-1}}{0.05\ mol\ dm^{-3}} = 3.02 \times 10^{-3}\ dm^3\ mol^{-1}\ s^{-1}$$

For the equilibrium

$$A \rightleftharpoons Y + Z$$

$$0.05 \times 0.1 \qquad 0.05 \times 0.9 \qquad 0.05 \times 0.9\ mol\ dm^{-3}$$

and the equilibrium constant k_1/k_{-1} is

$$\frac{(0.05 \times 0.9)^2}{0.05 \times 0.1} = 0.405\ mol\ dm^{-3}$$

Thus

$$k_{-1} = \frac{1.51 \times 10^{-4}\ s^{-1}}{0.405\ mol\ dm^{-3}} = 3.73 \times 10^{-4}\ dm^3\ mol^{-1}\ s^{-1}$$

The corresponding third-order rate constant is

$$\frac{3.73 \times 10^{-4}\ dm^3\ mol^{-1}\ s^{-1}}{0.05\ mol\ dm^{-3}} = 7.46 \times 10^{-3}\ dm^6\ mol^{-2}\ s^{-1}$$

38. If a_0 is the initial concentration of A and x the concentration of ions at equilibrium

$$\frac{dx}{dt} = k_1(a_0 - x) - k_{-1}x^2 \qquad (1)$$

At equilibrium

$$k_1(a_0 - x_e) - k_{-1}x_e^2 = 0 \qquad (2)$$

The deviation Δ_x from equilibrium, $x - x_e$ is given by

$$\frac{d\Delta x}{dt} = \frac{dx}{dt} = k_1(a_0 - x) - k_{-1}x^2 \tag{3}$$

Substraction of Eq. 2 leads to

$$\frac{d\Delta x}{dt} = -k_1\Delta x - k_{-1}(\Delta x)^2 - 2k_{-1}x_e\Delta x \tag{4}$$

Since Δx is very small the term in $(\Delta x)^2$ can be neglected:

$$\frac{d\Delta x}{dt} = -(k_1 + 2k_{-1}x_e)\Delta x \tag{5}$$

Integration gives

$$\ln \Delta x = -(k_1 + 2k_{-1}x_e)t + I \tag{6}$$

The boundary condition is $t = 0$, $\Delta x = (\Delta x)_0$, where $(\Delta x)_0$ is the initial value of Δx. Thus

$$\ln \frac{(\Delta x)_0}{\Delta x} = (k_1 + k_{-1}x_e)t \tag{7}$$

By definition, the relaxation time t^* is the time corresponding to

$$\frac{(\Delta x)_0}{\Delta x} = \mathrm{e} \tag{8}$$

Thus

$$t^* = \frac{1}{k_1 + k_{-1}x_e} \tag{9}$$

CHAPTER 10: WORKED SOLUTIONS
COMPOSITE REACTION MECHANISMS

1. $v = k_1[A][B]$

2. $v = k_2(k_1/k_{-1})^{1/2}[A]^{1/2}[B]$

3. $$v = \frac{k_1 k_2[A][B]}{k_{-1} + k_2[B]}$$

   (a) $$v = \frac{k_1 k_2}{k_{-1}}[A][B]$$

   (b) $v = k_1[A]$

4. $2A \rightarrow X$ (very slow)

   $X + 2B \rightarrow 2Y + 2Z$ (very fast)

5. Two simultaneous reactions

6. Two consecutive reactions

7. The steady-state equation for NO_3 is

   $$k_1[N_2O_5] - k_{-1}[NO_2][NO_3] - k_2[NO][NO_3] = 0$$

   Thus

   $$[NO_3] = \frac{k_1[N_2O_5]}{k_{-1}[NO_2] + k_2[NO]}$$

   The rate of consumption of N_2O_5 is

   $$v_{N_2O_5} = k_1[N_2O_5] - k_{-1}[NO_2][NO_3]$$
   $$= k_1[N_2O_5] - \frac{k_1 k_{-1}[N_2O_5]}{k_{-1}[NO_2] + k_2[NO]}$$
   $$= \frac{k_1 k_2[N_2O_5][NO]}{k_{-1}[NO_2] + k_2[NO]}$$

   Alternatively, we can note that $v_{N_2O_5} = v_{NO}$, from the stoichiometric equation, and the latter is $k_2[NO][NO_3]$.

8. The steady-state equation is

$$k_1[NO]^2 - k_{-1}[N_2O_2] - k_2[N_2O_2][O_2] = 0$$

and therefore

$$N_2O_2 = \frac{k_1[NO]^2}{k_{-1} + k_2[O_2]}$$

The rate is

$$v = v_{O_2} = k_2[N_2O_2][O_2] = \frac{k_1k_2[NO]^2[O_2]}{k_{-1} + k_2[O_2]}$$

This reduces to

$$v = \frac{k_1k_2[NO]^2[O_2]}{k_{-1}}$$

if $k_{-1} >> k_2[O_2]$.

9. The mechanism is

$$Cl_2 \xrightarrow{k_1} 2Cl \qquad \text{Initiation}$$

$$Cl + CH_4 \xrightarrow{k_2} HCl + CH_3$$

$$CH_3 + Cl_2 \xrightarrow{k_3} CH_3Cl + Cl$$

(Chain propagation)

$$2Cl \xrightarrow{k_{-1}} Cl_2$$

The steady-state equations are:

for Cl: $k_1[Cl_2] - k_2[Cl][CH_4] + k_3[CH_3][Cl_2] - k_{-1}[Cl_2]^2 = 0$

for CH_3: $k_2[Cl][CH_4] - k_3[CH_3][Cl_2] = 0$

Adding: $k_1[Cl_2] - k_{-1}[Cl]^2 = 0$

and therefore $[Cl] = \left(\frac{k_1}{k_{-1}}\right)^{1/2}[Cl_2]^{1/2}$

The rate of reaction is the rate of formation of HCl:

$$v = v_{HCl} = k_2[Cl][CH_4] = k_2\left(\frac{k_1}{k_{-1}}\right)^{1/2}[Cl_2]^{1/2}[CH_4]$$

10. The steady-state equation for O is

$$k_1[O_3]^2 - k_{-1}[O_3][O_2][O] - k_2[O][O_3] = 0$$

$$[O] = \frac{k_1[O_3]^2}{k_{-1}[O_3][O_2] + k_2[O_3]} = \frac{k_1[O_3]}{k_{-1}[O_2] + 2k_2}$$

$$v_{O_2} = k_1[O_3]^2 - k_{-1}[O_3][O_2][O] + 2k_2[O][O_3]$$

Subtraction of the steady-state equation gives

$$v_{O_2} = 3k_2[O][O_3]$$

$$= \frac{3k_1k_2[O_3]^2}{k_{-1}[O_2] + k_2}$$

The overall reaction is $2O_3 \rightarrow 3O_2$, and therefore

$3v_{O_3} = 2v_{O_2}$; thus

$$v_{O_3} = \frac{2k_1k_2[O_3]^2}{k_{-1}[O_2] + k_2}$$

$$= 2k_1[O_3]^2 \text{ in the absence of } O_2$$

11. The dissociation energy per molecule is

$$390.4 \times 10^3 \text{ J mol}^{-1}/6.022 \times 10^{23} \text{ mol}^{-1}$$

$$= 6.48 \times 10^{-19} \text{ J}$$

This corresponds to a frequency of

6.48×10^{-19} J/6.626×10^{-34} J s = 9.78×10^{14} s^{-1}

and to a wavelength of

$$\lambda = \frac{2.998 \times 10^{8} \text{ m s}^{-1}}{9.78 \times 10^{14} \text{ s}^{-1}} = 3.06 \times 10^{-7} \text{ m}$$

$$= 306 \text{ nm}$$

This is the maximum wavelength.

12. 440 μg of HI is $\frac{440 \times 10^{-6} \text{ g}}{127.9 \text{ g mol}} = 3.44 \times 10^{-6}$ mol

$$= 2.07 \times 10^{18} \text{ molecules}$$

207 nm corresponds to a frequency of

$$\frac{2.998 \times 10^{8} \text{ m s}^{-1}}{207 \times 10^{-9} \text{ m}} = 1.448 \times 10^{15} \text{ s}^{-1}$$

and to an energy of

6.626×10^{-34} J s x 1.448×10^{15} s^{-1} = 9.60×10^{-19} J

Therefore 1 J of radiation corresponds to

$1/9.60 \times 10^{-19} = 1.04 \times 10^{18}$ photons

One photon therefore decomposes 2.07/1.04

= 2 molecules

Possible mechanism:

$$HI + h\nu \rightarrow HI^*$$

$$HI^* + HI \rightarrow H_2 + I_2$$

13. The frequency of the radiation is

2.998×10^{8} m s^{-1}/253.7×10^{-9} m = 1.18×10^{15} s^{-1}

and its energy is

6.26×10^{-34} J s $\times 1.18 \times 10^{15}$ s^{-1} = 7.83×10^{-19} J

A 100 W lamp emits 100 J per second and therefore

$$\frac{100 \text{ J s}^{-1}}{7.83 \times 10^{-19} \text{ J}} = 1.28 \times 10^{20} \text{ photons per second}$$

This is

$$\frac{1.28 \times 10^{20} \text{ s}^{-1}}{6.022 \times 10^{23} \text{ mol}^{-1}} = 2.12 \times 10^{-4} \text{ moles of photons per second}$$

To decompose 0.01 mol requires

$$0.01/2.12 \times 10^{-4} = 47 \text{ seconds}$$

14. The lamp emits 2.12×10^{-4} moles of photons per second (see previous problem). In one hour it therefore emits

$$2.12 \times 10^{-4} \text{ mol s}^{-1} \times 3600 \text{ s} = 0.76 \text{ mol}$$

This is the amount of acetylene produced.

15. The steady-state equations are

For Cl: $2I_a - k_2[\text{Cl}][\text{CHCl}_3] + k_3[\text{CCl}_3][\text{Cl}_2] - k_4[\text{Cl}^2] = 0$

For CCl_3: $k_2[\text{Cl}][\text{CHCl}_3] - k_3[\text{CCl}_3][\text{Cl}_2] = 0$

Adding: $2I_a - k_4[\text{Cl}]^2 = 0$

$$[\text{Cl}] = \left(\frac{2I_a}{k_4}\right)^{1/2}$$

$$v = v_{\text{HCl}} = k_2[\text{Cl}][\text{CHCl}_3] = k_2\left(\frac{2}{k_4}\right)^{1/2} I_a^{1/2}[\text{CHCl}_3]$$

16. H^+ ions are most easily formed by the process

$$e^- + H_2O \rightarrow H^+ + OH + 2e^-$$

This is the sum of $H_2O \rightarrow H + OH$ and $e^- + H \rightarrow H^+ + 2e^-$, and its ΔH° is therefore

$$498.7 + 1312.2 = 1810.9 \text{ kJ mol}^{-1}$$
$$= 18.8 \text{ eV}$$

This is lower than the observed value of 19.5 eV, indicating that the system passes through a state of higher energy.

O^- ions are most easily formed by

$$e^- + H_2O \rightarrow 2H + O^-$$

which is the sum of $H_2O \rightarrow H + OH$, $OH \rightarrow H + O$ and $O + e^- \rightarrow O^-$.

Its ΔH° is therefore

$$498.7 - 428.2 - 213.4 = 713.5 \text{ kJ mol}^{-1} = 7.4 \text{ eV}$$

This is close to the observed appearance potential of 7.5 eV.

17. The concentration of H^+ ions in the solution is

$$\frac{3.2 \times 10^{-5}}{4.7 \times 10^{-2}} = 6.81 \times 10^{-4} \text{ mol dm}^{-3}$$

The dissociation constant is therefore

$$K_a = \frac{(6.81 \times 10^{-4})^2}{10^{-3}} = 4.64 \times 10^{-4} \text{ mol dm}^{-3}$$

18. The steady-state equations are:

For I: $k_1[I_2] - k_2[CH_3CHO][I] + k_4[CH_3][HI] - k_{-1}[I]^2 = 0$ (1)

For CH_3CO: $k_2[CH_3CHO][I] - k_3[CH_3CO] = 0$ (2)

For CH_3: $k_3[CH_3CO] - k_4[CH_3][HI] = 0$ (3)

(1) + (2) + (3) $k_1[I_2] - k_{-1}[I]^2 = 0$

Then

$$[I] = \left(\frac{k_1}{k_{-1}}\right)^{1/2} [I_2]^{1/2}$$

$$v = v_{CO} = k_3[CH_3CO] = k_2[CH_3CHO][I]$$

$$= k_2\left(\frac{k_1}{k_{-1}}\right)^{1/2} [I_2]^{1/2}[CH_3CHO]$$

19. The equations can be written as

$$\frac{1}{\rho}\log_{10}k = \frac{1}{\rho}\log_{10}k_0 + \sigma$$

and

$$\frac{1}{\rho'}\log_{10}K = \frac{1}{\rho'}\log_{10}K_0 + \sigma$$

Subtraction leads to

$$\frac{1}{\rho}\log_{10}k - \frac{1}{\rho'}\log_{10}K = \text{const}$$

and therefore

$$\log_{10}\frac{k^{1/\rho}}{K^{1/\rho'}} = \text{const}$$

or

$$\log_{10}\frac{k}{K^{\rho/\rho'}} = \text{const}$$

Thus,

$$k = GK^{\rho/\rho'} = GK^{\alpha}$$

where G and α are constants, the latter being equal to the ratio ρ/ρ'.

20. The HCl concentration remains unchanged throughout the experiment and the reaction is pseudo-first order with a rate constant of

$$2.80 \times 10^{-5}\ \text{dm}^3\ \text{mol}^{-1}\ \text{s}^{-1} \times 0.01\ \text{mol dm}^{-3}$$

$$= 2.80 \times 10^{-7}\ \text{s}^{-1}$$

The half life (Eq. 9.40) is

$$t_{1/2} = \frac{\ln 2}{2.80 \times 10^{-7}\ \text{s}^{-1}} = 2.48 \times 10^{6}\ \text{s}$$

21. Put $S = Co(NH_3)_5Cl^{2+}$

$X = Co(NH_3)_4(NH_2)Cl^{+}$

$Y = Co(NH_3)_4(NH_2)^{2+}$

Steady-state equations:

For X: $k_1[S][OH^-] - (k_{-1} + k_2)[X] = 0$

For Y: $k_2[X] - k_3[Y] = 0$

$$[X] = \frac{k_1[S][OH^-]}{k_{-1} + k_2}$$

$$[Y] = \frac{k_1(k_2/k_3)[S][OH^-]}{k_{-1} + k_2}$$

$$v = k_3[Y] = \frac{k_1k_2[S][OH^-]}{k_{-1} + k_2}$$

The dependence of v on [S] and $[OH^-]$ is independent of the relative magnitudes of the rate constants.

22.

$[S]/(10^{-3}$ mol $dm^{-3})$	v	$1/v$	$1/([S]/10^{-3}$ mol $dm^{-3})$	$v/([S]/10^{-3}$ mol $dm^{-3})$
0.4	2.41	0.415	2.50	6.025
0.6	3.33	0.300	1.667	5.55
1.0	4.78	0.209	1.000	4.78
1.5	6.17	0.162	0.667	4.11
2.0	7.41	0.135	0.500	3.705
3.0	8.70	0.115	0.333	2.90
4.0	9.52	0.105	0.250	2.38
5.0	10.5	0.095	0.200	2.10
10.0	12.5	0.080	0.100	1.25

From plots, K_m = 2.0 mmol dm^{-3}. See plots on page 173.

23.

$[S]/(10^{-6}$ mol $dm^{-3})$	$v/(10^{-6}$ mol dm^{-3} $s^{-1})$	$1/v(10^{-6}$ mol dm^{-3} $s^{-1})$	$1/([S]/10^{-6}$ mol $dm^{-3})$	$\frac{v/[S]}{10^{-3}s^{-1}}$
7.5	0.067	14.9	0.133	8.933
12.5	0.095	10.5	0.080	7.60
20.0	0.119	8.40	0.050	5.95
32.5	0.149	6.71	0.0308	4.58
62.5	0.185	5.41	0.0160	2.96
155.0	0.191	5.24	0.00645	1.232
320.0	0.195	5.13	0.00313	0.609

From graphs; K_m = 16 μmol dm^{-3}

V = 0.22 μmol dm^{-3}

See plots on page 174.

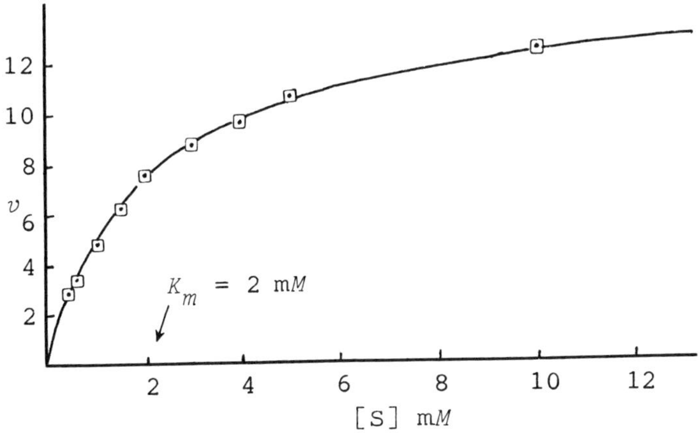
12
10
8
v
6
4
2
K_m = 2 mM
2
4
6
8
10
12
[S] mM

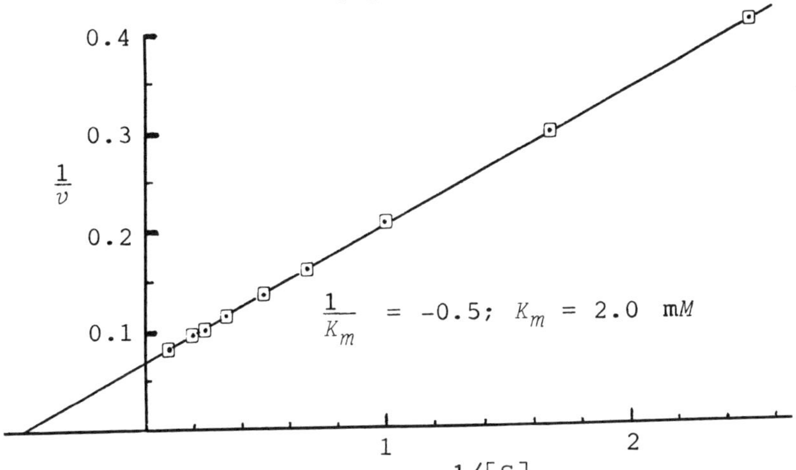
0.4
0.3
$\frac{1}{v}$
0.2
0.1
$\frac{1}{K_m}$ = -0.5; K_m = 2.0 mM
1
2
1/[S]

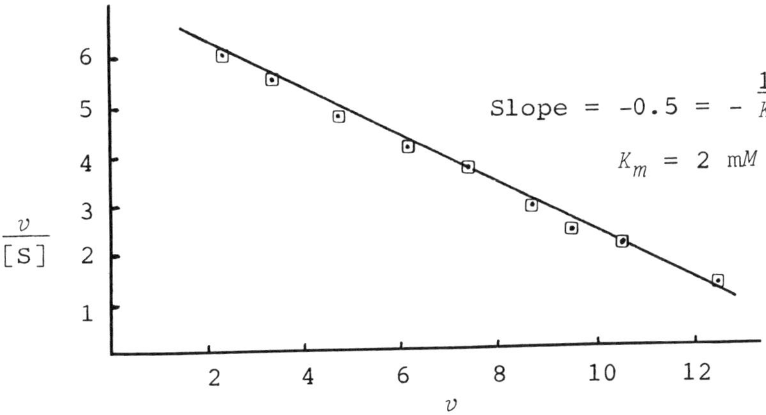
6
5
4
3
$\frac{v}{[S]}$
2
1
Slope = -0.5 = $-\frac{1}{K_m}$
K_m = 2 mM
2
4
6
8
10
12
v

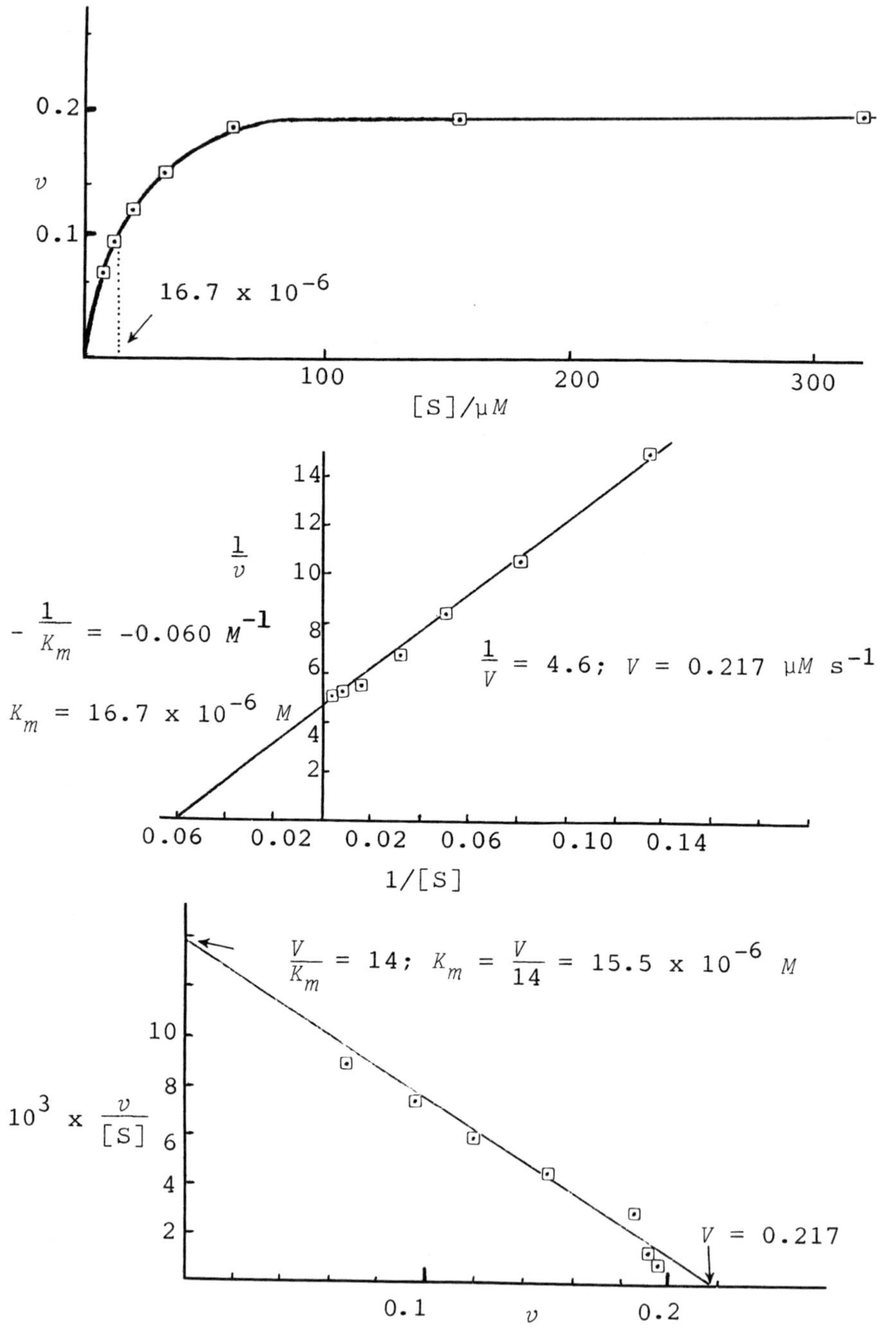

0.2
v
0.1
16.7 x 10^-6
100
200
300
[S]/μM
14
12
1/v
10
- 1/K_m = -0.060 M^-1
8
6
1/V = 4.6; V = 0.217 μM s^-1
K_m = 16.7 x 10^-6 M
4
2
0.06
0.02
0.02
0.06
0.10
0.14
1/[S]
V/K_m = 14; K_m = V/14 = 15.5 x 10^-6 M
10
8
10^3 x v/[S]
6
4
2
V = 0.217
0.1
v
0.2

24.

T/K	$V/10^{-6}$ mol $dm^{-3}s^{-1}$	$\frac{K_m}{10^{-4} mol\ dm^{-3}}$	$\frac{10^3}{(T/K)}$	$\log_{10}(V/10^{-6}$ mol $dm^{-3}s^{-1})$	$4 + \log_{10}$ $(K_m$/mol $dm^{-3})$
293	1.84	4.03	3.413	0.265	0.605
298	1.93	3.75	3.356	0.286	0.574
303	2.04	3.35	3.300	0.310	0.525
308	2.17	3.05	3.247	0.336	0.484

(a) Slope of plot of $\log_{10}V$ against $1/T$ = -427.5 K

(page 176).

E = 19.14 x 427.5 = 8180 J mol^{-1} = 8.18 kJ mol^{-1}

RT at 25°C = 2479 J mol^{-1} = 2.48 kJ mol^{-1}

$\Delta H^{\ddagger}$ = 8.18 - 2.48 = 5.7 kJ mol^{-1}

At 25.0°C, V = 1.93 x 10^{-6} dm^3 mol^{-1} s^{-1}

$[E]_0$ = 1.0 x 10^{-11} mol dm^3

$k_c = V/[E]_0$ = 1.93 x 10^5 s^{-1}

$k_c = \frac{\mathbf{k}T}{h} e^{-\Delta G^{\ddagger}/RT}$; $\frac{\mathbf{k}T}{h}$ at 25°C = 6.21 x 10^{12} s^{-1}

$e^{-\Delta G^{\ddagger}/RT}$ = 1.93 x 10^5/6.21 x 10^{12} = 3.108 x 10^{-8}

= $10^{-7.51}$

$\Delta G^{\ddagger}$ = 19.14 x 298.15 x 7.51 = 42 856 J mol^{-1}

= 42.9 kJ mol^{-1}

$\Delta S^{\ddagger} = \frac{5700 - 42\ 856}{298.15}$ = -124. J K^{-1} mol^{-1}

(b) Slope of $\log_{10}K_m$ against $1/T$ plot = 740 K

$\Delta H°$ (for dissociation) = -14 164 J mol^{-1}

= -14.2 kJ mol^{-1}

At 25.0°C, $\log_{10}$ $(K_m/10^{-4}$ $dm^{-3})$ = 0.57 - 4 = 3.43

$\Delta G°$ (for dissociation) = 19.14 x 298.15 x 3.43

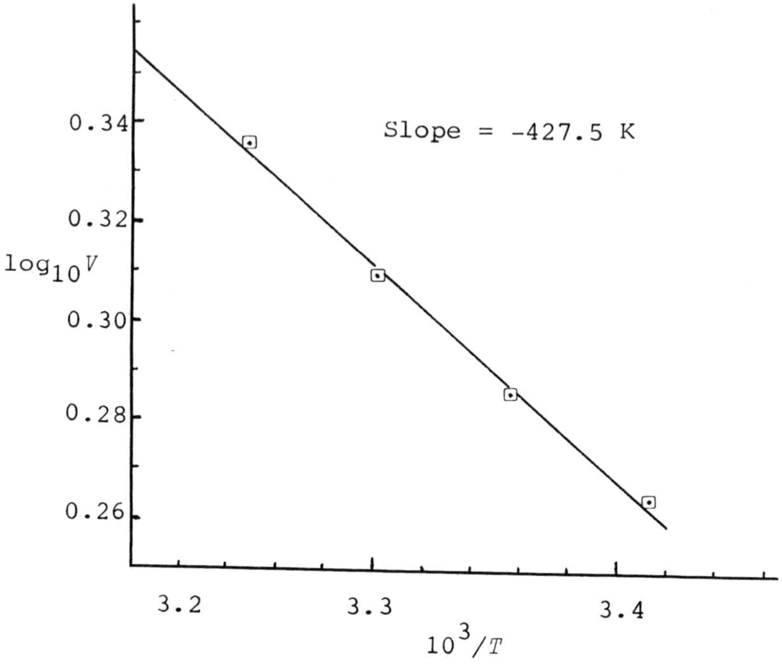

0.34
0.32
0.30
0.28
0.26
log10V
Slope = -427.5 K
3.2
3.3
3.4
10^3/T

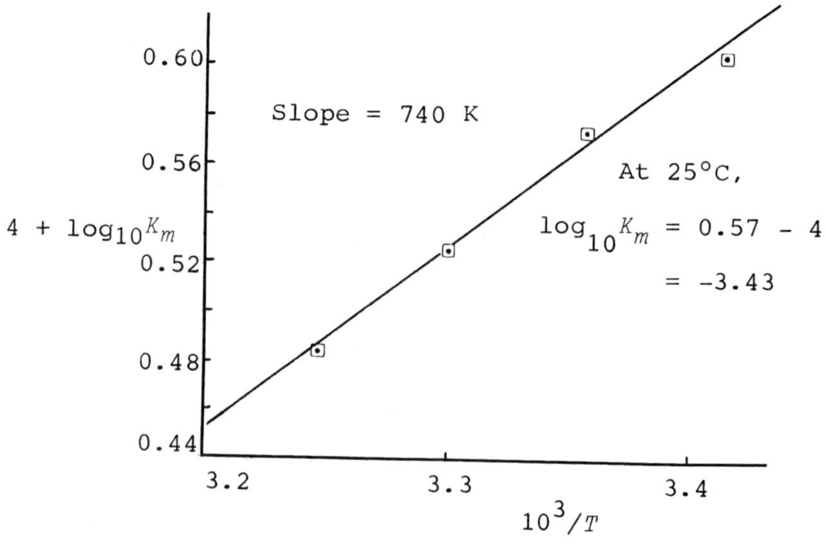

0.60
0.56
0.52
0.48
0.44
4 + log10Km
Slope = 740 K
At 25°C,
log10Km = 0.57 - 4
= -3.43
3.2
3.3
3.4
10^3/T

$$= 19\ 570\ \text{J mol}^{-1}$$

For association, $\Delta G^{\circ} = -19.6$ kJ mol^{-1};

$\Delta H^{\circ} = 14.2$ kJ mol^{-1}

$$\Delta S^{\circ} = \frac{14\ 164 + 19\ 570}{298.15} = 113\ \text{J K}^{-1}\ \text{mol}^{-1}$$

25.

$[S]/10^{-3}$ mol dm^{-3}	$V/10^{-5}$mol dm^{-3}s^{-1}	$\frac{1}{[S]/\text{mol dm}^{-3}}$	$1/v$
2	13	500	7692
4	20	250	5000
8	29	125	3448
12	33	83	3030
16	36	62.5	2778
20	38	50.0	2630

From a plot of $1/v$ against $1/[S]$, $V = 5.0 \times 10^{-4}$ mol dm^{-3} s^{-1}; $K_m = 5.8 \times 10^{-3}$ mol dm^{-3}

$$[E]_o = \frac{2\ \text{g dm}^{-3}}{50\ 000\ \text{g mol}^{-1}} = 4.0 \times 10^{-5}\ \text{mol dm}^{-3}$$

$$k_c = \frac{V}{[E]_o} = \frac{5.0 \times 10^{-4}}{4.0 \times 10^{-5}} = 12.5\ \text{s}^{-1}$$

26.

Temperature $T/°C$	T/K	$k_c/10^{-6}$ s^{-1}	$10^3/(T/K)$	$6 + \log_{10}(k_c/\text{s}^{-1})$
39.9	312.9	4.67	3.196	0.669
43.8	316.8	7.22	3.157	0.858
47.1	320.1	10.0	3.124	1.00
50.2	323.2	13.9	3.094	1.143

Slope of plot of $\log_{10}k_c$ against $1/T$ is

$= -4.73 \times 10^3$ K.

$E = 19.14 \times 4730 = 90\ 532$ J mol^{-1} = 90.5 kJ mol^{-1}

RT at 40°C = 8.314 x 313.15 = 2603 J mol^{-1}

$\Delta H^{\ddagger} = 90\ 500 - 2603 = 87\ 900$ mol^{-1} = 87.9 kJ mol^{-1}

$\frac{\mathbf{k}T}{h}$ at 40°C = 6.52×10^{12} s^{-1}; k_c at 40°C

$= 4.72 \times 10^{-6}$ s^{-1}

$$e^{-\Delta G^{\ddagger}/RT} = \frac{4.72 \times 10^{-6}}{6.52 \times 10^{12}} = 7.24 \times 10^{-19} = 10^{-18.14}$$

$\Delta G^{\ddagger} = 19.14 \times 313.15 \times 18.14 = 108\ 700$ J mol^{-1}

$= 108.7$ kJ mol^{-1}

$$\Delta S^{\ddagger} = \frac{87\ 900 - 108\ 700}{313.15} = -66.5 \text{ J K}^{-1} \text{ mol}^{-1}$$

27. Steady-state equation for EA:

$k_1[A][E] - k_{-1}[EA] - k_2[EA][B] = 0$

Steady-state equation for EY:

$k_2[EA][B] - k_3[EY] = 0$

$[E]_0 = [E] + [EA] + [EY]$

$$= [EA]\left\{\frac{k_{-1} + k_2[B]}{k_1[A]} + 1 + \frac{k_2[B]}{k_3}\right\}$$

$v = k_2[EA][B]$

$$= \frac{k_2[B][E]_0}{\dfrac{k_{-1} + k_2[B]}{k_1[A]} + 1 + \dfrac{k_2[B]}{k_3}}$$

$$= \frac{k_1 k_2 k_3[E]_0[A][B]}{k_{-1}k_3 + k_1k_3[A] + k_2k_3[B] + k_1k_2[A][B]}$$

28. $k_1[E][S] - (k_{-1} + k_2)[ES] = 0$

$$k_i[E][I] - k_{-i}[EI] = 0$$

$$[E]_0 = [E] + [ES] + [EI]$$

$$[E]_0 = [ES]\left\{\frac{k_{-1} + k_2}{k_1[S]} + 1 + \frac{k_i[I](k_{-1} + k_2)}{k_i k_1[S]}\right\}$$

$$v = k_2[ES] = \frac{k_2[E]_0}{\dfrac{k_{-1} + k_2}{k_1[S]} + 1 + \dfrac{k_i[I](k_{-1} + k_2)}{k_{-i}k_1[S]}}$$

$$= \frac{k_2[E]_0[S]}{\dfrac{k_{-1} + k_2}{k_1} + [S] + \dfrac{k_i(k_{-1} + k_2)}{k_{-i}k_1}[I]} = \frac{k_2[E]_0[S]}{K_m\left\{1 + \dfrac{[I]}{K_i}\right\} + [S]}$$

$$\varepsilon = \frac{v_0 - v}{v_0} = 1 - \frac{v}{v_0}$$

$$= 1 - \frac{k_2[E]_0[S]}{K_m\left\{1 + \dfrac{[I]}{K_i}\right\} + [S]} \cdot \frac{K_m + [S]}{k_2[E]_0[S]}$$

$$= 1 - \frac{K_m + [S]}{K_m\left\{1 + \dfrac{[I]}{K_i}\right\} + [S]}$$

$$= \frac{\dfrac{K_m}{K_i}[I]}{K_m\left\{1 + \dfrac{[I]}{K_i}\right\} + [S]}$$

29. $$E + S \underset{k_{-1}}{\overset{k_1}{\rightleftharpoons}} ES \xrightarrow[\curvearrowright\, Y]{k_2} ES' \xrightarrow{k_3} E + Z$$

$$k_1[E][S] - (k_{-1} + k_2)[ES] = 0$$

$$k_2[ES] - k_3[ES'] = 0$$

$$[ES]_0 = [E] + [ES] + [ES']$$

$$= [ES]\left\{\frac{k_{-1} + k_2}{k_1[S]} + 1 + \frac{k_2}{k_3}\right\}$$

$$v = k_2[ES] = \frac{k_2[E]_0}{\dfrac{k_{-1} + k_2}{k_1[S]} + 1 + \dfrac{k_2}{k_3}}$$

$$= \frac{\dfrac{k_2k_3}{k_2 + k_3}[E]_0[S]}{\dfrac{k_{-1} + k_2}{k_1}\,\dfrac{k_3}{k_2 + k_3} + [S]}$$

When [S] is large

$$v = \frac{k_1k_2}{k_{-1} + k_2}[E]_0$$

and the catalytic constant is therefore

$$k_c = \frac{k_1k_2}{k_{-1} + k_2}$$

K_m is the first term in the denominator of the rate equation:

$$K_m = \frac{k_{-1} + k_2}{k_1}\,\frac{k_3}{k_2 + k_3}$$

30. The steady-state equation is

$$k_1[A][B] - (k_{-1} + k_2)[X] = 0$$

and the rate is

$$v = \frac{k_1 k_2}{k_{-1} + k_2} = [A][B]$$

and the rate constant is $k = k_1 k_2/(k_{-1} + k_2)$.

$$\ln k = \ln k_1 + \ln k_2 - \ln(k_{-1} + k_2)$$

$$\frac{d\ln k}{dT} = \frac{d\ln k_1}{dT} + \frac{1}{k_2}\frac{dk_2}{dT} - \frac{1}{k_{-1} + k_2}\frac{d(k_{-1} + k_2)}{dT}$$

$$= \frac{d\ln k_1}{dT} + \frac{k_{-1}}{k_{-1} + k_2}\frac{d\ln k_2}{dT} - \frac{k_{-1}}{k_{-1} + k_2}\frac{d\ln k_1}{dT}$$

$$E \equiv RT^2 \frac{d\ln k}{dT}; E_1 \equiv RT^2 \frac{d\ln k_1}{dT}; E_{-1} \equiv RT^2 \frac{d\ln k_{-1}}{dT}$$

$$E_2 \equiv RT^2 \frac{d\ln k_2}{dT}$$

$$E = E_1 + \frac{k_{-1}}{k_{-1} + k_2} E_2 - \frac{k_{-1}}{k_{-1} + k_2} E_{-1}$$

$$= \frac{k_2 E_1 + k_{-1}(E_1 + E_2 - E_{-1})}{k_{-1} + k_2}$$

31. $$v = \frac{A_c\, e^{-E_c/RT}}{Be^{-\Delta H^\circ/RT} + [A]}$$

If $[A] >> Be^{-\Delta H^\circ/RT}$ $\quad E = E_c$

If $[A] << Be^{-\Delta H^\circ/RT}$ $\quad E = E_c - \Delta H^\circ$

If ΔH° is positive and is greater than E_c, E can become negative. With ΔH° positive $B\exp(-\Delta H^\circ/RT)$ increases with increasing temperature, and at sufficiently high temperatures E is therefore negative. The rate therefore rises with T at lower temperature, passes through a maximum, and then decreases.

32. The steady-state equation is

$$k_1[A]^2 - k_{-1}[A][A^*] - k_2[A^*] = 0$$

Therefore

$$[A^*] = \frac{k_1[A]^2}{k_{-1}[A] + k_2}$$

The rate is

$$v = k_2[A^*] = \frac{k_1k_2[A]^2}{k_{-1}[A] + k_2}$$

At high pressures $k_{-1}[A] >> k_2$ and thus

$$v = \frac{k_1k_2}{k_{-1}}[A] \qquad \text{First order}$$

At low pressures $k_{-1}[A] << k_2$ and then

$$v = k_1[A]^2 \qquad \text{Second order}$$

33. Steady-state equation for EA:

$$k_1[E][A] - k_{-1}[EA] - k_2[EA][B] = 0$$

Steady-state equation for EAB

$$k_2[EA][B] - (k_{-2} + k_3)[EAB] = 0$$

$$[\mathrm{E}]_0 = [\mathrm{E}] + [\mathrm{EA}] + [\mathrm{EAB}]$$

$$= [\mathrm{EAB}]\left\{\frac{k_{-1} + k_2[\mathrm{B}]}{k_1[\mathrm{A}]} \cdot \frac{k_{-2} + k_3}{k_2[\mathrm{B}]} + \frac{k_{-2} + k_3}{k_2[\mathrm{B}]} + 1\right\}$$

$$v = k_2[\mathrm{EAB}] = \frac{k_2[\mathrm{E}]_0}{\dfrac{k_{-1} + k_2[\mathrm{B}]}{k_1[\mathrm{A}]} \cdot \dfrac{k_{-2} + k_3}{k_2[\mathrm{B}]} + \dfrac{k_{-2} + k_3}{k_2[\mathrm{B}]} + 1}$$

$$= \frac{k_1 k_2 k_3 [\mathrm{E}]_0[\mathrm{A}][\mathrm{B}]}{k_{-1}(k_{-2} + k_3) + k_1(k_{-2} + k_3)[\mathrm{A}] + k_2(k_{-2} + k_3)[\mathrm{B}] + k_1 k_2[\mathrm{A}][\mathrm{B}]}$$

34. With the boundary condition $t = 0$, $c = c_0$ the differential equation integrates as follows:

$$-\int \frac{dc}{c^2} = k\,dt$$

$$\frac{1}{c} = kt + I, \text{ and } I = \frac{1}{c_0}$$

Thus

$$\frac{1}{c} - \frac{1}{c_0} = kt = \frac{c_0 - c}{c\,c_0}$$

The fraction of functional groups

$$f = \frac{c_0 - c}{c_0}$$

Elimination of c gives

$$\frac{f}{1 - f} = c_0kt$$

or

$$f = \frac{c_0kt}{1 + c_0kt}$$

CHAPTER 11: WORKED SOLUTIONS
QUANTUM MECHANICS AND ATOMIC STRUCTURE

1. (a) The frequency is calculated from $c = \lambda\nu$ where c is the speed of light (2.998×10^8 m s^{-1}):

$$\nu = \frac{c}{\lambda} = \frac{2.998 \times 10^8 \text{ m s}^{-1}}{325 \times 10^{-9} \text{ m}} = 9.22 \times 10^{14} \text{ s}^{-1}$$

(b) $\lambda = 325 \times 10^{-9}$ m $= 3250 \times 10^{-8}$ cm

$$\bar{\nu} = \frac{1}{\lambda} = 3.07 \times 10^4 \text{ cm}^{-1}$$

(c) $\varepsilon = h\nu = 6.626 \times 10^{-34}$ (J s) 9.22×10^{-14} s^{-1}

$= 6.085 \times 10^{-19}$ J

In molar units:

6.022×10^{23} mol^{-1} x 6.085×10^{-19} J

$= 3.66 \times 10^5$ J mol^{-1} = 366 kJ mol^{-1}

$E = h\nu = 6.085 \times 10^{-19}$ J x 6.2420×10^{18} eV/J

$= 3.80$ eV

(d) The momentum is calculated by the use of Eq. 11.56:

$$p = \frac{h}{\lambda} = \frac{6.626 \times 10^{-34} \text{ J s}}{3.25 \times 10^{-7} \text{ m}} = 2.039 \times 10^{-27} \text{ kg m s}^{-1}$$

2. (a) $$\lambda = c/\nu = \frac{2.998 \times 10^8 \text{ m s}^{-1}}{1.965 \times 10^8 \text{ s}^{-1}} = 1.526 \text{ m}$$

(b) $E = h\nu = 6.626 \times 10^{-34}$ J s x 196.5×10^6 s^{-1}

$= 1.302 \times 10^{-25}$ J

$= 8.127 \times 10^{-7}$ eV $= 0.078$ J mol^{-1}

(c) $$p = \frac{h}{\lambda} = \frac{6.626 \times 10^{-34} \text{ J s}}{1.526 \text{ m}} = 4.34 \times 10^{-34} \text{ kg m s}^{-1}$$

3. $$\nu_1 = \frac{c}{\lambda_1} = \frac{2.998 \times 10^8 \text{ m s}^{-1}}{766.494 \times 10^{-9} \text{ m}} = 3.9113 \times 10^{14} \text{ s}^{-1}$$

$$\nu_2 = \frac{c}{\lambda_2} = \frac{2.998 \times 10^8 \text{ m s}^{-1}}{769.901 \times 10^{-9} \text{ m}} = 3.8940 \times 10^{14} \text{ s}^{-1}$$

$$\nu_1 - \nu_2 = 0.0173 \times 10^{14}\ s^{-1} = 1.73 \times 10^{12}\ s^{-1}$$

4. The form of the equation is (compare Eq. 11.6)

$$y = A \sin(\omega t + \delta)$$

The angular frequency ω is $(3\pi/5)$ rad s^{-1}

From Eq. 11.4, $\omega = (2\pi \text{ rad})\nu$ and therefore

$$\nu = \frac{\omega}{2\pi \text{ rad}} = \frac{(3\pi/5) \text{ rad } s^{-1}}{2\pi \text{ rad}} = 0.3\ s^{-1}$$

5. (a) The period, $\tau = 1/\nu$ is related to the force constant of the spring by Eq. 11.15.

$$\text{period} = \frac{1}{\nu} = 2\pi \frac{m}{k_h}$$

Solving for k_h:

$$k_h = \frac{4\pi^2 m}{\tau^2} = \frac{4\pi^2 (0.2 \text{ kg})}{(3.0 \text{ s})^2} = 0.88 \text{ N m}^{-1}$$

(b) To obtain the velocity, differentiate the equation of motion for the harmonic oscillator (Eq. 11.6). Thus

$$v = \frac{dy}{dx} = \omega A \cos(\omega\tau + \delta)$$

Since $|\cos \theta| \leq 1$ for all θ, it follows that the maximum velocity is

$$v_{max} = \omega A = \sqrt{\frac{k_h}{m}}\, A = \sqrt{\frac{0.88 \text{ N/m}}{0.2 \text{ kg}}} \times 0.01\text{m}$$

$$= 0.021\text{m s}^{-1}$$

6. The low frequency limit can be obtained by use of the series expression

$$e^x = 1 + x + \frac{x^2}{2!} + \cdots$$

When x is small we may approximate e^x by $1 + x$. Thus when $h\nu \ll \mathbf{k}T$

$$\overline{E} = \frac{h\nu}{1 + \frac{h\nu}{\mathbf{k}T} - 1} = \mathbf{k}T$$

Electromagnetic waves are transverse waves having two degrees of freedom. This value $\mathbf{k}T$ is divided between these two degrees of freedom.

7. The energy of each photon is

$$h\nu = \frac{hc}{\lambda} = \frac{6.626 \times 10^{-34} \text{ J s} \times 2.998 \times 10^{8} \text{ m s}^{-1}}{550 \times 10^{-9} \text{ m}}$$

$$= 3.61 \times 10^{-19} \text{ J}$$

The lamp emits 50 J per second and the number of photons is therefore

$$\frac{50}{3.61 \times 10^{-19}} = 1.39 \times 10^{20}$$

The momentum of each is

$$p = \frac{h}{\lambda} = \frac{6.626 \times 10^{-34} \text{ J s}}{550 \times 10^{-9} \text{ m}} = 1.204 \times 10^{-27} \text{ kg m s}^{-1}$$

8. Eq. 11.37, $h\nu = \frac{1}{2}mu^2 + w$, gives $h\nu_o = w$ when the kinetic energy $\frac{1}{2}mu^2$ is zero; ν_o is the threshold frequency. Thus,

$$h\nu_o = w = 6.626 \times 10^{-34} \text{ (J s) } 43.9 \times 10^{13} \text{ s}^{-1}$$

$$= 2.91 \times 10^{-19} \text{ J}$$

or, in eV, $w = 2.91 \times 10^{-19}\ \text{J}/1.602 \times 10^{-19}\ \text{J/eV}$

$= 1.82\ \text{eV}$

With the more modern value

$$w = 6.626 \times 10^{-34}\ \text{J s} \times 5.5 \times 10^{13}\ \text{s}^{-1}$$

$$= 3.64 \times 10^{-20}\ \text{J} = 0.23\ \text{eV}$$

9. (a) Using Eq. 11.56, $\lambda = \frac{h}{mu}$, we have

$$\lambda = \frac{6.626 \times 10^{-34}\ \text{J s}}{9.11 \times 10^{-31}\ \text{kg} \times 6.0 \times 10^{7}\ \text{m s}^{-1}}$$

$$= 1.21 \times 10^{-11}\ \text{m} = 12.1\ \text{pm}$$

(b) $$\lambda = \frac{h}{mu} = \frac{6.626 \times 10^{-34}\ \text{J s}}{\dfrac{32.0\ \frac{\text{g}}{\text{mol}}\ \text{mol}}{6.022 \times 10^{23}}\ \dfrac{1\ \text{kg}}{1000\ \text{g}}\ 425\ \text{m s}^{-1}}$$

$$= 2.93 \times 10^{-11}\ \text{m} = 29.3\ \text{pm}$$

(c) $$\lambda = \frac{h}{mu} = \frac{6.626 \times 10^{-34}\ \text{J s}}{\dfrac{4.0\ \frac{\text{g}}{\text{mol}} \cdot \text{mol}}{6.022 \times 10^{23}}\ \dfrac{1\ \text{kg}}{1000\ \text{g}}\ 1.5 \times 10^{7}\ \text{m s}^{-1}}$$

$$= 6.65 \times 10^{-15}\ \text{m} = 6.65\ \text{fm}$$

(d) $$\lambda = \frac{h}{mu} = \frac{6.626 \times 10^{-34}}{9.11 \times 10^{-31}(2.818 \times 10^{8})} = 2.58 \times 10^{-12}\ \text{m} = 2.58\ \text{fm}$$

10. From Eq. 11.61, the product of the uncertainties in position and velocity, is

$$\Delta q \Delta u \approx \frac{h}{4\pi m} = \frac{6.626 \times 10^{-34}\ \text{J s}\ (= \text{kg m}^2\ \text{s}^{-1})}{4 \times 3.14 \times 6 \times 10^{-16}\ \text{kg}}$$

$$= 10^{-19}\ \text{m}^2\ \text{s}^{-1}$$

Therefore, since $\Delta q = 10^{-9}$ m,

$$\Delta u = 10^{-10} \text{ m s}^{-1}$$

With this uncertainty in velocity, the position of the particle one second later would be uncertain to within 2×10^{-10} m, or 0.2 nm. This is only 0.2% of the diameter of the particle, and the uncertainty principle therefore does not present a serious problem for particles of this magnitude. For particles of molecular sizes, the uncertainty is much greater.

11. (a) The kinetic energy of the electron is

$$E_k = 10 \times 1.602 \times 10^{-19} \text{ V C} = 1.602 \times 10^{-18} \text{ J}$$

This energy is $\frac{1}{2}mu^2$ and therefore

$$u = \sqrt{\frac{2 \times 1.602 \times 10^{-18} \text{ J}}{9.11 \times 10^{-31} \text{ kg}}} = 1.875 \times 10^6 \text{ m s}^{-1}$$

Then, from Eq. 11.56,

$$\lambda = \frac{6.626 \times 10^{-34} \text{ J s}}{9.11 \times 10^{-31} \text{ kg} \times 1.875 \times 10^6 \text{ m s}^{-1}}$$

$$= 3.88 \times 10^{-10} \text{ m} = 388 \text{ pm}$$

(b) $$u = \sqrt{\frac{2 \times 10^3 \times 1.602 \times 10^{-19} \text{ J}}{9.11 \times 10^{-31} \text{ kg}}} = 1.875 \times 10^7 \text{ m s}^{-1}$$

$$\lambda = \frac{6.626 \times 10^{-34}}{9.11 \times 10^{-31} \times 1.875 \times 10^7} = 3.88 \times 10^{-11} \text{ m}$$

$$= 38.8 \text{ pm}$$

(c) $$u = \sqrt{\frac{2 \times 10^6 \times 1.602 \times 10^{-19}\ \text{J}}{9.11 \times 10^{-31}\ \text{kg}}} = 5.93 \times 10^8\ \text{m s}^{-1}$$

$$\lambda = \frac{6.626 \times 10^{-34}}{9.11 \times 10^{-31} \times 5.93 \times 10^8} = 1.23 \times 10^{-12}\ \text{m}$$

$$= 1.23\ \text{pm}$$

12. A particle of mass m has a frequency of

$$\nu = \tfrac{1}{2}mu^2/h$$

and a de Broglie wavelength (Eq. 11.56) of

$$\lambda = \frac{h}{mu}$$

Elimination of u between these equations gives

$$\nu = \frac{h}{2m}\left(\frac{1}{\lambda^2}\right)$$

The group velocity is thus

$$v_g = \frac{d\nu}{d(1/\lambda)} = 2\cdot\frac{h}{2m}\cdot\frac{1}{\lambda} = \frac{h}{m\lambda} = u$$

13. (a) $N^2\int(\psi_1 + \psi_2)^2\ d\tau = 1$

$N^2[\int\psi_1{}^2 d\tau + \int\psi_2{}^2 d\tau + 2\int\psi_1\psi_2 d\tau] = 1$

The first 2 integrals are equal to unity because ψ_1 and ψ_2 are normalized, and $2\int\psi_1\psi_2 d\tau = 0$ because the wave functions are orthogonal. The result is $2N^2 = 1$ or $N = \frac{1}{\sqrt{2}}$. The normalized wave function is thus $\frac{1}{\sqrt{2}}(\psi_1 + \psi_2)$

(b) $N^2\int(\psi_1 - \psi_2)^2 d\tau = 1$

$$N^2[\int\psi_1{}^2 d\tau + \int\psi_2{}^2 d\tau + 2\int\psi_1\psi_2 d\tau] = 1$$

Because of normalization the first two integrals are equal to 1. The integral $\int\psi_1\psi_2 d\tau = 0$ by the orthogonality condition. Thus, $N^2(2) = 1$ or $N = \frac{1}{\sqrt{2}}$ and the normalized wave function is $\frac{1}{\sqrt{2}}(\psi_1 - \psi_2)$.

(c) $N^2\int(\psi_1 + \psi_2 + \psi_3)^2 d\tau = 1$

$$N^2[\int\psi_1{}^2 d\tau + 2\int\psi_1\psi_2 d\tau + 2\int\psi_1\psi_3 d\tau + \int\psi_2{}^2 d\tau +$$

$$2\int\psi_2\psi_3 d\tau + \int\psi_3{}^2] = 1$$

By the normalization and orthogonality conditions,

$$N^2[1 + 2\ (0) + 2\ (0) + 1 + 2\ (0) + 1] = 1$$

$$3\ N^2 = 1$$

$$N = \frac{1}{\sqrt{3}}$$

and the wavefunction is $\frac{1}{\sqrt{3}}(\psi_1 + \psi_2 + \psi_3)$

(d) $N^2\int(\psi_1 - \frac{1}{\sqrt{2}}\ \psi_2 + \frac{\sqrt{3}}{\sqrt{2}}\ \psi_3)\ d\tau = 1$

$$N^2[\int\psi_1{}^2 d\tau + \frac{1}{2}\int\psi_2{}^2 d\tau + \frac{3}{2}\int\psi_3{}^2 d\tau - \frac{\sqrt{2}}{\sqrt{2}}\int\psi_1\psi_2 d\tau -$$

$$\frac{2\sqrt{3}}{\sqrt{2}}\int\psi_1\psi_2 d\tau - \frac{2\sqrt{3}}{\sqrt{2}}\int\psi_1\psi_3 d\tau - \int\frac{2\sqrt{3}}{2}\psi_2\psi_3 d\tau] = 1$$

By the normalization and orthogonality conditions

$$N^2[1 + \frac{1}{2} + \frac{3}{2} - 0 - 0 - 0] = 1$$

$$3N^2 = 1$$

$N = \frac{1}{\sqrt{3}}$ and the wave function is

$$\frac{1}{\sqrt{3}}(\psi_1 - \frac{1}{\sqrt{2}}\psi_2 + \frac{\sqrt{3}}{\sqrt{2}}\psi_3) = \frac{1}{\sqrt{3}}\psi_1 - \frac{1}{\sqrt{6}}\psi_2 + \frac{1}{\sqrt{2}}\psi_3$$

14. If Ae^{-ax} is an eigenfunction of $\frac{d^2}{dx^2}$, operation on Ae^{-ax} twice by d/dx will give the original function multiplied by a constant:

$$\frac{d}{dx}(Ae^{-ax}) = -Aae^{-ax}$$

$$\frac{d}{dx}(-Aae^{-ax}) = Aa^2e^{-ax}$$

Therefore, it is an eigenfunction, with eigenvalue a^2.

15. For the wavefunction to be single-valued $\sin[m_l(\phi + 2\pi)]$ must equal $\sin m_l\phi$.

The former function is

$$\sin[m_l\phi + 2\pi m_l] = \sin m_l\phi \cos 2\pi m_l + \sin 2\pi m_l \cos m_l\phi$$

For this to equal $\sin m_l\phi$

$\cos 2\pi m_l$ must be 1

and $\sin 2\pi m_l$ must be 0.

This can only occur if m_l is an integer.

16. As in Eq. 11.83 we may express Ψ as

$$\Psi(x,y,z,t) = \psi(x,y,z)e^{-2\pi iEt/h}$$

Then

$$\Psi^* = \psi^*(x,y,z)\ e^{2\pi i E t/h}$$

and

$$\Psi^*\Psi = \psi^*\psi$$

which is independent of time.

17. The operator for p_x (Table 11.1) is

$$\frac{h}{2\pi i}\frac{\partial}{\partial x}$$

If $\phi(x)$ and $\psi(x)$ are any two functions, the Hermitian condition is

$$\int_{-\infty}^{\infty} \phi^* \frac{h}{2\pi i}\frac{d\psi}{dx}\,dx = \int_{-\infty}^{\infty} \psi \left(\frac{h}{2\pi i}\right)^* \frac{d\phi^*}{dx}\,dx \qquad (1)$$

Integration by parts of the left-hand-side gives

$$\int_{-\infty}^{\infty} \phi^* \frac{h}{2\pi i}\frac{d\psi}{dx}\,dx = \frac{h}{2\pi i}\left[\phi^*\psi\right]\Big|_{-\infty}^{\infty} - \int_{-\infty}^{\infty} \psi\, \frac{h}{2\pi i} \cdot \frac{d\phi^*}{dx}\,dx \qquad (2)$$

The second term is equal to

$$\int_{-\infty}^{\infty} \psi\left(\frac{h}{2\pi i}\right)^* \frac{d\phi^*}{dx}\,dx$$

which is the right-hand-side of Eq. 1. The operator is therefore Hermitian.

18. (a) $\frac{dk}{dx} = 0 = 0 \times k$; k is an eigenfunction with the eigenvalue 0.

(b) $\frac{d\ kx^2}{dx} = 2kx$; kx^2 is not an eigenfunction

(c) $\frac{d \sin kx}{dx} = k \cos kx$; $\sin kx$ is not an eigenfunction

(d) $\frac{d\ e^{kx}}{dx} = k\ e^{kx}$; e^{kx} is an eigenfunction with the eigenvalue k.

(e) $\frac{d\ e^{kx^2}}{dx} = 2kx\ e^{kx^2}$; e^{kx^2} is not an eigenfunction

(f) $\frac{d\ e^{ikx}}{dx} = ik\ e^{ikx}$; e^{ikx} is an eigenfunction with the eigenvalue ik

19. The lowest energy is given by Eq. 11.147 with $n_1 = n_2 = n_3 = 1$.

(a) If $a = 1.0 \times 10^{-11}$ m,

$$E = \frac{3(6.626 \times 10^{-34}\ \text{J s})^2}{8 \times 9.1 \times 10^{-31}\ \text{kg} \times (10^{-11}\ \text{m})^2}$$
$$= 1.81 \times 10^{-15}\ \text{J}$$
$$= 1.13 \times 10^{4}\ \text{eV}$$

(b) If $a = 10^{-15}$ m

$$E = \frac{3(6.626 \times 10^{-34}\ \text{J s})^2}{8 \times 9.1 \times 10^{-31}\ \text{kg} \times (10^{-15}\ \text{m})^2}$$
$$= 1.81 \times 10^{-7}\ \text{J}$$
$$= 1.13 \times 10^{12}\ \text{eV}$$

The latter energy is so large that the electron would not remain in the nucleus, but would be emitted as a β particle.

20. (a) The normalization condition is

$$\int_a^b \psi\ \psi^* dx = 1 = \int_a^b \frac{A^2}{x^2}\, dx$$

$$= A^2 \left[-\frac{1}{x}\right]_a^b = A^2 \left[-\frac{1}{b} + \frac{1}{a}\right]$$

Therefore

$$A^2 = \frac{ab}{b - a} \quad \text{and } A = \sqrt{\frac{ab}{a - b}}$$

(b) The average value of x is

$$\langle x \rangle = \int_a^b \psi^*\ x\ \psi\ dx$$

$$= A^2 \int_a^b \frac{1}{x}\, dx = A^2\, [\ln x]\Big|_a^b$$

$$= \frac{ab}{a-b} \ln \frac{b}{a}$$

21. The energy of the nth level is, from Eq. 11.145

$$E_n = \frac{n^2 h^2}{8ma^2} = \frac{n^2 (6.626 \times 10^{-34}\ \text{J s})^2}{8 \times 9.11 \times 10^{-31}\ \text{kg} \times (10^{-9}\ \text{m})^2}$$

$$= 6.024 \times 10^{-20}\ n^2\ \text{J}$$

$$= 0.376\ n^2\ \text{eV}$$

At 10 eV, $n = \sqrt{\dfrac{10}{0.376}} = 5.17$

Thus levels 1 and 5 have energies less than 10 eV.

At 100 eV, $n = \sqrt{\dfrac{100}{0.376}} = 16.31$

Therefore levels 6 and 16, i.e. 11 levels, have energies between 10 and 100 eV.

22. The unnormalized solution for a particle in a one-dimensional box (Eq. 11.140) is

$$\psi_n = \sin \frac{n\pi x}{a}$$

The momentum operator is $\frac{h}{2\pi i}\frac{d}{dx}$:

$$\frac{h}{2\pi i}\frac{d\psi_n}{dx} = \frac{h}{2\psi i}\frac{n\pi}{a}\cos\frac{n\pi x}{a}$$

Since the result is not a constant multiplied by ψ_n, ψ_n is not an eigenfunction of the momentum operator. This conclusion is related to the Heisenberg uncertainty principle; the position and momentum operators do not commute, there are no common eigenfunctions, and the two properties can not be measured simultaneously and precisely. However, the eigenfunction ψ_n, like any other function, can be expressed as a linear combination of the set of momentum operators (compare Eqs. 11.117 to 11.120). The physical significance of this is that the function ψ_n corresponds to a wave train of particular momentum being reflected at the walls of the box and giving rise to a wave train in the opposite direction.

23. Frequency of vibration,

$$\nu_0 = 2.998 \times 10^{10}\ \text{cm s}^{-1} \times 2360\ \text{cm}^{-1}$$
$$= 7.075 \times 10^{13}\ \text{s}^{-1}$$

$$h\nu_0 = 7.075 \times 10^{13} \times 6.626 \times 10^{-34} = 4.69 \times 10^{-20} \text{ J}$$

$$\text{Zero-point energy} = \tfrac{1}{2}h\nu_0 = 2.34 \times 10^{-20} \text{ J}$$

Energy at $v = 1$ is $\tfrac{3}{2}h\nu_0 = 7.03 \times 10^{-20}$ J

24. The energy is related to the angular momentum by Eq. 11.206:

$$E = \frac{L^2}{2I}$$

The Hamiltonian operator is therefore

$$\hat{H} = \frac{1}{2I}\left(\frac{h}{2\pi i}\right)^2 \frac{\partial^2}{\partial \phi^2}$$

$$= -\frac{h^2}{8\pi^2 I}\frac{\partial^2}{\partial \phi^2}$$

25. The energy required to remove the electron from the lowest energy level in hydrogen ($n_1 = 1$) to infinity ($n_2 = \infty$) is the ionization potential. Using Eq. 11.38 and the value of the Rydberg constant, we have

$$\bar{\nu} = 1.0968 \times 10^7 \text{ (m}^{-1}\text{)}\left(\frac{1}{1^2} - \frac{1}{\infty^2}\right) = 1.0968 \times 10^7 \text{ m}^{-1}$$

The energy required is

$$E = hc\bar{\nu} = 6.626 \times 10^{-34} \text{ (J s) } 2.998 \times 10^8 \text{ (m s}^{-1}\text{)} \cdot$$
$$1.0968 \times 10^{17} \text{ m}^{-1} = 2.18 \times 10^{-18} \text{ J}$$

$$= \frac{2.18 \times 10^{-18} \text{ J}}{1.602 \times 10^{-10} \text{ J/eV}} = 13.6 \text{ eV}$$

26. From Eq. 11.40, with $n = 1$,

$$u = \frac{h}{2\pi m r}$$

and by Eq. 11.44, $r = a_o = 52.92$ pm.

Therefore

$$u = \frac{6.626 \times 10^{-34} \text{ J s}}{2\pi \times 9.1 \times 10^{-31} \text{ kg} \times 5.29 \times 10^{-11} \text{ m}}$$

$$= 2.19 \times 10^{6} \text{ m s}^{-1}$$

By Eq. 11.56,

$$\lambda = \frac{h}{mu} = \frac{6.626 \times 10^{-34} \text{ J s}}{9.1 \times 10^{-31} \text{ kg} \times 2.1\text{g} \times 10^{6} \text{ m s}^{-1}}$$

$$= 3.32 \times 10^{-10} \text{ m} = 332 \text{ pm}$$

From Eq. 11.40

$$u = \frac{nh}{2\pi m r}$$

and therefore, since for $Z = 1$, $r = n^2 a_0$

$$\lambda = \frac{h}{mu} = \frac{2\pi r}{n} = 2\pi a_0 n$$

The circumference of the orbit is $2\pi n^2 a_0$, and the ratio of the circumference to λ is therefore n.

27. $$\bar{\nu} = \frac{1}{\lambda} = 1.0968 \times 10^{7} \text{ m}^{-1} \left(\frac{1}{4^2} - \frac{1}{5^2}\right)$$

$$= 2.469 \times 10^{5} \text{ m}^{-1}$$

$$\lambda = \frac{1}{\bar{\nu}} = 4.05 \times 10^{-6} \text{ m}$$

This is in the infrared region of the electromagnetic spectrum.

$$E = hc\bar{\nu} = 6.626 \times 10^{-34}\ (\text{J s})\ 2.998 \times 10^{8}\ (\text{s m}^{-1}) \cdot$$
$$2.469 \times 10^{5}\ \text{m}^{-1} = 4.90 \times 10^{-20}\ \text{J}$$

28. From Eq. 11.49, with $Z = 1$ and $n = 2$,

$$E_p = -\frac{e^2}{4\pi\varepsilon_0 4a_0}$$
$$= -\frac{(1.602 \times 10^{-19}\ \text{C})^2}{16\pi(8.854 \times 10^{-12}\ \text{C}^2\ \text{N}^{-1}\ \text{m}^{-2})(5.292 \times 10^{-11}\ \text{m})}$$
$$= -\ 10.9 \times 10^{-18}\ \text{J}$$

In atomic units of $e^2/4\pi\varepsilon_0 a_0$,

$$E_p = -\ 0.25\ \text{a.u.}$$

29. Problems 29 and 30 are conveniently worked out with reference to Problem 25, where the ionization energy of H was calculated to be 13.6 eV. According to Eq. 11.50 the first ionization energy is proportional to Z_{eff}^2/n^2, where n is the quantum number of the most easily removed electron. For H, $Z_{eff} = 1$ and $n = 1$, so that the ionization energy is

$$I/\text{eV} = 13.6\ \frac{Z_{eff}^2}{n^2}$$

For Li, $n = 2$ and therefore

$$5.39 = 13.2\ \frac{Z_{eff}^2}{4}$$

Thus

$$Z_{eff} = 1.27$$

30. Similarily, for Na, where $n = 3$,

$$5.14 = 13.2 \frac{Z^2_{eff}}{9}$$

and

$$Z_{eff} = 1.87$$

31. (a) The nuclear charge of Cl is 17. We subtract

0.30 for the other 3s electron,

5 x 0.35 for the five 3p electrons,

8 x 0.85 for the eight 2s and 2p electrons,

2 x 1.00 for the two 1s electrons.

Thus, $\sigma = 10.85$

and $Z_{eff} = 17 - 10.85 = 6.15$

(b) The nuclear charge of P is 15. We subtract

2 x 0.30 for the 3s electron,

2 x 0.35 for the other 3p electrons,

8 x 0.85 for the eight 2s and 2p electrons,

2 x 1.00 for the two 1s electrons.

Thus, $\sigma = 10.1$

and $Z_{eff} = 15 - 10.1 = 4.9$

(c) The nuclear charge of K is 19. We subtract

8 x 0.85 for the 3s and 3p electrons,

10 x 1.00 for the 1s, 2s and 2p electrons,

Thus, $\sigma = 16.8$

and $Z_{eff} = 19 - 16.8 = 2.2$

32. (a) The basic equation is

$$-\frac{h^2}{8\pi^2 m}\nabla^2\psi + E_p(x,y,z)\psi = E\psi$$

(b) The potential energy E_p can be set equal to zero inside the box, and therefore

$$\frac{\partial^2\psi}{\partial x^2} + \frac{\partial^2\psi}{\partial y^2}\frac{\partial^2\psi}{\partial z^2} = -\frac{8\pi^2 mE}{h^2}\psi$$

The energy E is separated into its component parts: $E = E_x + E_y + E_z$, and ψ is factored into the functions $X(x)$, $Y(y)$, and $Z(z)$. These values are substituted and the whole expression is divided by XYZ:

$$\frac{1}{XYZ}\frac{\partial^2 XYZ}{\partial x^2} + \frac{1}{XYZ}\frac{\partial^2 XYZ}{\partial y^2} + \frac{1}{XYZ}\frac{\partial^2 XYZ}{\partial z^2} = -\frac{8\pi^2 mE}{h^2}$$

In the first term, Y and Z do not depend on x and can be brought outside the derivative. We thus obtain three equations; one is

$$-\frac{1}{X}\frac{\partial^2 X}{\partial x^2} = \frac{8\pi^2 mE_x}{h^2}$$

and there are equivalent equations for Y and Z.

(c) The above equations are identical with Eq. 11.124. The solutions are of the form of Eq. 11.140:

$$\psi_n = C\sin\frac{n\pi x}{a}$$

and normalization gives $C = \frac{2}{a}$

(d) The total energy is

$$E = E_x + E_y + E_z = \frac{n_x^2 h^2}{8ma^2} + \frac{n_y^2 h^2}{8mb^2} + \frac{n_z^2 h^2}{8mc^2}$$

If $a = b = c$

$$E = (n_x^2 + n_y^2 + n_z^2) \frac{h^2}{8ma^2}$$

33. From Eq. 11.148

$$\mu_H = \frac{m_e m_H}{m_e + m_H} = \frac{9.1095 \times 1.6727 \times 10^{-58}}{9.1095 \times 10^{-31} + 1.6727 \times 10^{-27}}$$

$$= \frac{15.2374 \times 10^{-58}}{1.6736 \times 10^{-27}} = 9.1046 \times 10^{-31} \text{ kg}$$

$$\mu_D = \frac{m_e m_D}{m_e + m_D} = \frac{9.1095 \times 3.3434 \times 10^{-58}}{9.1095 \times 10^{-31} + 3.3434 \times 10^{-27}}$$

$$= \frac{30.4567 \times 10^{-58}}{3.3443 \times 10^{-27}} = 9.1070 \times 10^{-31} \text{ kg}$$

(a) By Eq. 11.45 the Bohr radius is inversely proportional to μ, and is therefore slightly smaller for D than for H.

The energies are inversely proportional to the Bohr radii (Eq. 11.50) and are therefore slightly greater for D than for H. The frequencies of the transitions are therefore slightly greater for D and the wavelengths slightly shorter.

(b) The wavelengths are in the inverse ratio of the reduced masses, and are therefore in the ratio

$$\frac{\lambda(D)}{\lambda(H)} = \frac{9.1046}{9.1070}$$

The wavelength of the line in the spectrum of D is therefore

$$\frac{9.1046}{9.1070} \times 656.47 = 656.30 \text{ nm}$$

34. From the expression for ψ in Table 11.5,

$$\psi^*\psi = \frac{1}{\pi a_0^3} e^{-2r/a_0}$$

Multiplying by the volume element in spherical coordinates we have

$$\frac{1}{\pi a_0^3} e^{-2r/a_0} r^2 dr \sin\theta \, d\theta \, d\phi$$

Integration of θ from 0 to π and ϕ from 0 to 2π gives

$$\int_0^\pi \sin\theta \, d\theta = -\cos\theta \Big|_0^\pi = -[(-1) - (1)] = 2$$

and

$$\int_0^\pi d\phi = 2\pi$$

so that

$$\iint \frac{1}{\pi a_0^3} e^{-2r/a_0} r^2 dr \sin\theta \, d\theta \, d\phi$$

$$= \frac{4\pi}{\pi a_0^3} r^2 e^{-2r/a_0} dr$$

$$= \frac{4}{a_0^3} r^2 e^{-2r/a_0} dr$$

35. Since the function in Tables 11.2 and 11.3 are real, $\Theta\Phi\Theta^*\Phi^*$ can be replaced by $\Theta^2\Phi^2$. The functions we require are

$$\Theta_{10}\Phi_0 = \frac{\sqrt{6}}{2} \cos\theta \frac{1}{\sqrt{2\pi}}$$

$$\Theta_{11}\Phi_1 = \frac{\sqrt{3}}{2} \sin\theta \frac{1}{\sqrt{\pi}} \cos\phi$$

$$\Theta_{1-1}\Phi_{-1} = \frac{\sqrt{3}}{2} \sin\theta \frac{1}{\sqrt{\pi}} \sin\phi$$

The sum of their squares is

$$\frac{3}{4\pi} \cos^2\theta + \frac{3}{4\pi} \sin^2\theta \cos^2\phi + \frac{3}{4\pi} \sin^2\theta \sin^2\phi$$

$$= \frac{3}{4\pi} \cos^2\theta + \frac{3}{4\pi} \sin^2\theta (\cos^2\phi + \sin^2\phi)$$

$$= \frac{3}{4\pi} \quad (\text{since } \cos^2\theta + \sin^2\theta = 1)$$

This is independent of θ and ϕ.

36. $\Delta q \Delta p = \dfrac{h}{4\pi}$

and $\dfrac{p^2}{2m} = E$

$\Delta p = \sqrt{2m\Delta E}$

$$\Delta E = \frac{(\Delta p)^2}{2m} = \frac{h^2}{32\pi^2 m(\Delta q)^2}$$

(a) For a cube of sides 10 pm = 1.0×10^{-11} m

$$\Delta E = \frac{(6.626 \times 10^{-34} \text{ J s})^2}{32\pi^2 \times 9.11 \times 10^{-32} \text{ kg } (1.0 \times 10^{-11} \text{ m})^2}$$

$$= 1.53 \times 10^{-17} \text{ J} = 95.3 \text{ eV}$$

(b) For a cube of sides 1 fm = 10^{-15} m

$$\Delta E = \frac{(6.626 \times 10^{-34} \text{ J s})^2}{32\pi^2 \times 9.11 \times 10^{-31} \text{ kg } (1.0 \times 10^{-15} \text{ m})^2}$$

$$= 1.53 \times 10^{-9} \text{ J} = 9.53 \times 10^9 \text{ eV}$$

These uncertainties are considerably smaller than the energies calculated for the particle in a box. They are in fact smaller by the factor

$$\frac{3/8}{1/32\pi^2} = 12\pi^2 = 118.4$$

37. Wave functions for levels m and n are given by (Eq. 11.144)

$$\psi_m = \sin\frac{m\pi x}{a} \text{ and } \psi_n = \sin\frac{n\pi x}{a}$$

(the normalization coefficients are unnecessary for proving orthogonality). Then

$$\int_0^a \psi_m\psi_n = \int_0^a \sin\frac{m\pi x}{a}\sin\frac{n\pi x}{a}\,dx$$

Put $y = \pi x/a$; the limits are now π and 0:

Then $dy = (\pi/a)dx$ and the integral becomes

$$\frac{a}{\pi}\int_0^{\pi} \sin my \sin ny \, dy$$

This is a standard integral and has the value

$$\frac{a}{\pi}\left[\frac{\sin(m-n)y}{2(m-n)} - \frac{\sin(m+n)y}{2(m+n)}\right]_0^{\pi}$$

At the lower limit ($y = 0$), both terms are zero since $\sin 0^{\circ} = 0$. At the upper limit both terms are zero since if m and n are integers $\sin(m-n)$ and $\sin(m+n)$ are zero. The integral is thus zero and the wavefunctions are orthogonal.

CHAPTER 12: WORKED SOLUTIONS
THE CHEMICAL BOND

1. The resultant energy is

$$E_p/\text{kJ mol}^{-1} = -\frac{137.2}{r/\text{nm}} + \frac{0.0975}{(r/\text{nm})^6}$$

$$\frac{d(E_p/\text{kJ mol}^{-1})}{d(r/\text{nm})} = \frac{137.2}{(r/\text{nm})^2} - \frac{6 \times 0.0875}{(r/\text{nm})^7}$$

This is zero when $r = r_0$; thus

$$(r_0/\text{nm})^5 = \frac{6 \times 0.0975}{137.2} = 4.264 \times 10^{-3}$$

$$r_0 = 0.336 \text{ nm} = 336 \text{ pm}$$

When $r = 0.336$ nm,

$$E_p = -\frac{137.2}{0.336} + \frac{0.0975}{0.336^6} = -340.5 \text{ kJ mol}^{-1}$$

2. If the bond were completely ionic the dipole moment would be

$$\mu = (1.602 \times 10^{-19} \text{ C})\ (239 \times 10^{-12} \text{ m})$$

$$= 3.83 \times 10^{-29} \text{ C m}$$

The percentage ionic character is thus

$$\frac{2.09}{3.83} \times 100 = 55$$

3. (a) $$E_{\text{ionic}} = D(\text{LiH}) - [D(\text{Li}_2)\ D(\text{H}_2)]^{1/2}$$

$$= 243 - (113 \times 435)^{1/2} = 22 \text{ kJ mol}^{-1}$$

$$\therefore\ |\chi_{\text{Li}} - \chi_{\text{H}}| = \frac{\sqrt{22}}{10} = 0.47$$

Thus the estimated dipole moment is 0.47 D.

(b) If the molecule were completely ionic the dipole

moment would be

$$4.8 \times (1.26 + 0.36) = 7.78 \text{ D}$$

$$\% \text{ ionic character} = \frac{0.47}{7.78} \times 100 = 6.0\%$$

4. | | | |
|---|---|---|
| $BeCl_2$ | (no lone pairs) | linear |
| SF_6 | (no lone pairs) | octahedral |
| H_3O^+ | (one lone pair) | triangular-pyramid |
| NH_4^+ | (no lone pair) | tetrahedral |
| PCl_6^- | (no lone pair) | octahedral |
| AlF_6^{3-} | (no lone pair) | octahedral |
| PO_4^{3-} | (no lone pair) | tetrahedral |
| CO_2 | (no lone pair) | linear |
| SO_2 | (one lone pair) | bent |
| NH_3^{2+} | (no lone pair) | planar |
| CO_3^{2-} | (no lone pair) | planar |
| NO_3^- | (no lone pairs) | planar |

5. HCl: $\mu_{ionic} = 1.602 \times 10^{-19} \text{ C} \times 127 \times 10^{-12} \text{ m}$

$= 2.034 \times 10^{-29} \text{ C m}$

$$\% \text{ ionic character} = \frac{3.60 \times 10^{-30} \times 100}{2.034 \times 10^{-29}} = 17.7\%$$

HBr: $\mu_{ionic} = 1.602 \times 10^{-19} \text{ C} \times 141 \times 10^{-12} \text{ m}$

$= 2.26 \times 10^{-29} \text{ C m}$

$$\% \text{ ionic character} = \frac{2.67 \times 10^{-30} \times 100}{2.26 \times 10^{-29}} = 11.8\%$$

HI: $\mu_{ionic} = 1.602 \times 10^{-19}$ C x 160×10^{-12} m

$= 2.56 \times 10^{-29}$ C m

$$\% \text{ ionic character} = \frac{1.40 \times 10^{-30} \times 100}{2.56 \times 10^{-29}} = 5.5\%$$

CO: $\mu_{ionic} = 1.602 \times 10^{-19}$ C x 113×10^{-12} m

$= 1.81 \times 10^{-29}$ C m

$$\% \text{ ionic character} = \frac{0.33 \times 10^{-30} \times 100}{1.81 \times 10^{-29}} = 1.8\%$$

6. The MO diagrams are based on Figures 12.20a and 12.22. Then:

No. of electrons		Configuration	Bond Order	Number of Unpaired Electrons
B_2	10	$(1s\sigma_g)^2(1s\sigma_u^*)^2(2s\sigma_g)^2(2s\sigma_u^*)^2(2p\sigma_g)^2$	1	0
CO	14	$(1s\sigma)^2(1s\sigma^*)^2(2s\sigma)^2(2s\sigma^*)^2(2p\sigma)^2(2p\pi)^4$	3	0
BN	12	$(1s\sigma)^2(1s\sigma^*)^2(2s\sigma)^2(2s\sigma^*)^2(2p\sigma)^2(2p\pi)^2$	2	2
BN^{2-}	14	Same as CO	3	0
BO	13	$(1s\sigma)^2(1s\sigma^*)^2(2s\sigma)^2(2s\sigma^*)^2(2p\sigma)^2(2p\pi)^3$	2.5	1
BF		Same as CO	3	0
OF	17	$(1s\sigma)^2(1s\sigma^*)^2(2s\sigma)^2(2s\sigma^*)^2(2p\sigma)^2(2p\pi)^4(2p\pi^*)^3$	1.5	1
OF^-	18	$(1s\sigma)^2(1s\sigma^*)^2(2s\sigma)^2(2s\sigma^*)^2(2p\sigma)^2(2p\pi)^4(2p\pi^*)^4$	1	0
OF^+	16	$(1s\sigma)^2(1s\sigma^*)^2(2s\sigma)^2(2s\sigma^*)^2(2p\sigma)^2(2p\pi)^4(2p\pi^*)^2$	2	1

Species with no unpaired electrons are diamagnetic, others are paramagnetic.

7. $\underline{N_2}$. See Fig. 12.21a. An added electron goes into the antibonding $2p\pi^*$ level; the bond is weakened. An electron removed leaves from the $2p\sigma_g$ level; the bond is weakened.

$\underline{O_2}$. See Fig. 12.21b. Added electron goes into antibonding $2p\pi_g^*$ level; the bond is weakened. An electron removed leaves from the $2p\pi_g^*$ level; the bond is strengthened.

$\underline{C_2}$. The electronic configuration is

$$(1s\sigma_g)^2(1s\sigma_u^*)^2(2s\sigma_g)^2(2s\sigma_u^*)^2(2p\sigma_g)^2(2p\pi_u)^2$$

Removal of a $2p_u$ electron will weaken the bond; addition will strengthen it.

$\underline{F_2}$. The configuration is

$$(1s\sigma_g)^2(1s\sigma_u^*)^2(2s\sigma_g)^2(2s\sigma_u^*)^2(2p\pi_u)^4(2p\sigma_g)^2(2p\pi_g^*)^4$$

Removal of a $2p\pi_g^*$ electron will strengthen the bond; an added electron will go into the $2p\sigma_u^*$ level and weaken it.

$\underline{CN}$. The configuration is

$$(1s\sigma)^2(1s\sigma^*)^2(2s\sigma)^2(2s\sigma^*)^2(2p\sigma)^2(2p\pi)^3$$

Removal of a $2p\pi$ electron will weaken the bond; an added electron will go into the $2p\pi$ level and strengthen it.

$\underline{NO}$. The configuration is

$(1s\sigma)^2(1s\sigma^*)^2(2s\sigma)^2(2s\sigma^*)^2(2p\sigma)^2(2p\pi)^4\ 2p\pi^*$

Removal of a $2p\pi^*$ electron will strengthen the bond; an added electron will go into the $2p\pi^*$ level and weaken it.

Thus,

(a) Addition of an electron will strengthen C_2 and CN, and will weaken N_2, O_2, F_2, NO.

(b) Removal of an electron will strengthen O_2, F_2, and NO, and will weaken N_2, C_2, and CN.

8. We are given that

$$\int (1s_A)^2\, d\tau = 1 \quad \text{and} \quad \int (1s_B)^2\, d\tau = 1$$

We are required to prove that

$$\int (1s_A + 1s_B)(1s_A - 1s_B)\, d\tau = 0$$

(The 1s wavefunctions are real so that they are the same as their complex conjugates). The integral is

$$\int (1s_A)^2\, d\tau - \int (1s_B)^2\, d\tau$$

which is zero since each term is unity.

9.

Object	Symmetry Elements	Point Group
Equilateral triangle	E, C_3, $3C_2$, σ_h, $3\sigma_v$, $2S_3$	D_{3h}
Isosceles triangle	E, C_2, $2\sigma_v$	C_{2v}
Cylinder	E, C_∞ ∞C_2, $\infty\sigma_v$, σ_h, i	$D_{\infty h}$

10.

Molecule	Symmetry Elements	Point Group
$CHCl_3$	$E,\ C_3,\ 3\sigma_v$	C_{3v}
CH_2Cl_2	$E,\ C_2,\ 2\sigma$	C_{2v}
Naphthalene	$E,\ C_2,\ 2C_2,\ 3\sigma,\ i$	D_{2h}
NO_2	$E,\ C_2,\ 2\sigma_v$	C_{2v}
Cyclopropane	$E,\ C_3,\ 3C_2,\ 3\sigma_v,\ \sigma_h$	D_{3h}
CO_3^{2-}	$E,\ C_3,\ 3C_2,\ 3\sigma_v,\ \sigma_h$	D_{3h}
C_2H_2	$E,\ C_\infty,\ \infty C_2,\ \infty\sigma_v,\ \sigma_h,\ i$	$D_{\infty h}$

11. H_2O_2 belongs to the C_2 point group, and has neither a plane of symmetry nor a center of inversion. It therefore can exist in two enantiomeric forms. However, optical activity has not been detected because of the rapid interconversion of the two forms.

12. (a) H_2; point group $D_{\infty h}$; σ_g^+

(b) H_2O; point group C_{2v}; a_1, b_2, a_1.

(c) CO_2; point group $D_{\infty h}$; σ_g^+, σ_u^+

(d) BF_3; point group D_{3h}; a_1'

(e) NH_3; point group C_{3v}; a_1

(f) HCN; point group $C_{\infty v}$; σ^+

13. (a) $\int t_3 t_3^* \, d\tau = \int t_3^2 \, d\tau$ (since the functions are real)

$$= \frac{1}{4}\int (s - p_x + p_y - p_z)^2 \, d\tau$$

$$= \frac{1}{4}\int (s^2 + p_x{}^2 + p_y{}^2 + p_z{}^2 - 2sp_x + 2sp_y{}^2 + \ldots)\, d\tau$$

All cross terms are zero, and the remainder are unity; the integral is thus

$$\frac{1}{4}(1 + 1 + 1 + 1) = 1$$

(b) $\int t_2 t_4{}^* \, d\tau = \int t_2 t_4 \, d\tau$ (since the functions are real)

$$= \frac{1}{4}\int (s + p_x - p_y - p_z)(s - p_x - p_y + p_z)\, d\tau$$

$$= \frac{1}{4}\int (s^2 - p_x^2 + p_y^2 - p_z^2 + \text{cross terms which are zero})\, d\tau$$

$$= \frac{1}{4}(1 - 1 + 1 - 1) = 0$$

14. $\int \psi_1^2 \, d\tau = \int (as + bp_x)^2 \, d\tau = \int (a^2 s^2 + b^2 p_x^2 + 2absp_x)\, d\tau = a^2 + b^2 = 1$ (1)

$$\int \psi_2^2 \, d\tau = \int \left(as - \tfrac{1}{2}bp_x + \tfrac{\sqrt{3}}{2}cp_y\right)^2 d\tau$$

$$= a^2 + \frac{b^2}{4} + \frac{3}{4}c^2 = 1 \quad (2)$$

$$\int \psi_3^2 \, d\tau = \int \left(as - \tfrac{1}{2}bp_x - \tfrac{\sqrt{3}}{2}cp_y\right)^2 d\tau$$

$$= a^2 + \frac{b^2}{4} + \frac{3}{4}c^2 = 1 \quad (3)$$

$$\int \psi_1\psi_2 \, d\tau = \int (as + bp_x)\left(as - \tfrac{1}{2}bp_x + \tfrac{\sqrt{3}}{2}cp_y\right) d\tau$$

$$= a^2 - \tfrac{1}{2}b^2 = 0 \quad (4)$$

From (1) and (4) $\frac{3}{2}b^2 = 1; \quad b = \sqrt{\frac{2}{3}}$

$$a^2 = \frac{1}{3}; \quad a = \frac{1}{\sqrt{3}}$$

From (3) $\frac{1}{3} + \frac{1}{6} + \frac{3}{4}c^2 = 1; \; c^2 = \frac{2}{3}; \; c = \sqrt{\frac{2}{3}}$

[This has only used the fact that ψ_1 and ψ_2 are orthogonal; it is easily confirmed that ψ_3 is also orthogonal to ψ_1 and ψ_2]. The normalized wave-functions are thus

$$\psi_1 = \frac{1}{\sqrt{3}}\,\mathrm{s} + \frac{2}{\sqrt{3}}\,\mathrm{p}_x$$

$$\psi_2 = \frac{1}{\sqrt{3}}\,\mathrm{s} - \frac{1}{\sqrt{6}}\,\mathrm{p}_x + \frac{1}{\sqrt{2}}\,\mathrm{p}_y$$

$$\psi_3 = \frac{1}{\sqrt{3}}\,\mathrm{s} - \frac{1}{\sqrt{6}}\,\mathrm{p}_x - \frac{1}{\sqrt{2}}\,\mathrm{p}_y$$

15. The diagram shows the Z axis and the contributions along that axis from s and p_z orbitals. The wave-functions are thus

$$\psi_1 = a\mathrm{s} + b\mathrm{p}_z$$

$$\psi_2 = a\mathrm{s} - b\mathrm{p}_z$$

The normalization condition for ψ_1 is

$$\int \psi_1^2 \, d\tau = (a\mathrm{s} + b\mathrm{p}_z)^2 \, d\tau = a^2 + b^2 = 1$$

The orthogonality condition is

$$\int \psi_1\psi_2 \, d\tau = (a\mathrm{s} + b\mathrm{p}_z)(a\mathrm{s} - b\mathrm{p}_z) \, d\tau$$

$$= a^2 - b^2 = 0$$

Thus,

$$2b^2 = 1; \; b = \frac{1}{\sqrt{2}}; \; a = \frac{1}{\sqrt{2}}$$

The wavefunctions are therefore

$$\psi_1 = \frac{1}{\sqrt{2}}(s + p_z)$$

$$\psi_2 = \frac{1}{\sqrt{2}}(s - p_z).$$

16.

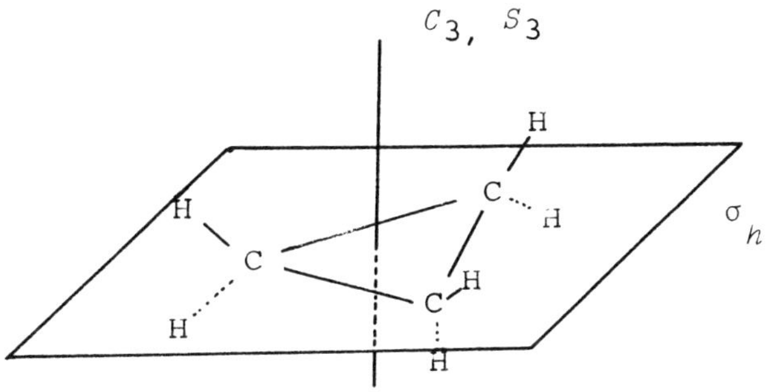

The C_3 axis is also the S_3 axis:

$$S_3 = C_3\sigma_h$$

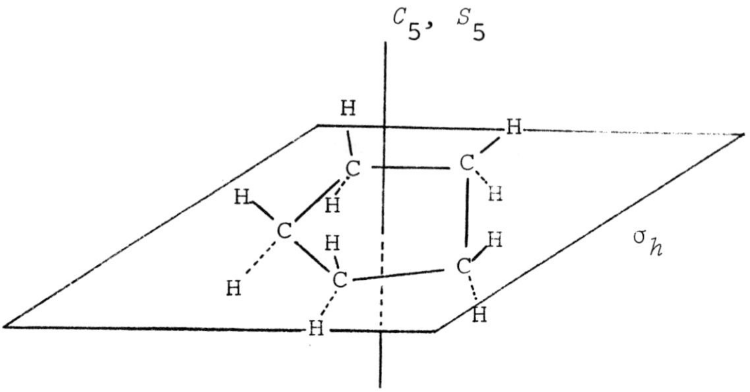

The C_5 axis is the S_5 axis:

$$S_5 = C_5\sigma_h$$

The point groups

$$C_{2h},\ D_{2h},\ D_{\infty h},\ \text{and}\ O_h$$

have centers of symmetry. The groups

$$D_{3h}\ \text{and}\ T_d$$

have axes of improper rotation. The groups in Table 12.3 for which there can be a dipole moment are the remaining ones, viz.

$$C_1,\ C_2,\ C_{2v},\ C_{3v},\ \text{and}\ C_{\infty v}.$$

CHAPTER 13: WORKED SOLUTIONS
CHEMICAL SPECTROSCOPY

1. $I_0/I = 78/55$

   Absorbance = $\log_{10}$ (78/55) = $\log_{10}$ 1.418 = 0.152

   Transmittance = I/I_0 = 55/78 = 0.705

   Molar absorption coefficient, $\varepsilon = \dfrac{0.152}{0.1 \text{ mol dm}^{-3} \times 0.5 \text{ cm}}$

   $= 3.04 \text{ dm}^3 \text{ mol}^{-1} \text{ cm}^{-1}$

2. Absorbance, $A = \log_{10} \dfrac{1}{0.87} = 0.060$

   Concentration, $c = \dfrac{9.5 \text{ g dm}^{-3}}{18\ 800} = 5.05 \times 10^{-4} \text{ mol dm}^{-3}$

   $\varepsilon = \dfrac{0.0605}{5.05 \times 10^{-4} \times 10.0} = 12.0 \text{ dm}^3 \text{ mol}^{-1} \text{ cm}^{-1}$

3. $c = \dfrac{0.155}{532 \times 1.0} = 2.91 \times 10^{-4} \text{ mol dm}^{-3}$

4. The absorbance at 340 gives [NADH]:

   $[\text{NADH}] = \dfrac{0.215}{6.22 \times 10^3 \times 1} = 3.46 \times 10^{-5} \text{ mol dm}^{-3}$

   At 260 nm this concentration will give an absorbance

   of

   $1.8 \times 10^4 \times 3.46 \times 10^{-5} = 0.623.$

   The remaining absorbance, 0.850 - 0.623 = 0.227 is due

   to the NAD^+:

   $[\text{NAD}^+] = \dfrac{0.227}{1.80 \times 10^4} = 1.04 \times 10^{-5} \text{ mol dm}^{-3}$

5. $A = \log_{10} \dfrac{1}{0.280} = 0.553$

$$\varepsilon = \frac{0.553}{0.01 \text{ mol dm}^{-3} \times 0.20 \text{ cm}} = 276 \text{ dm}^3 \text{ mol}^{-1} \text{ cm}^{-1}$$

In a cell 1 cm thick

$A = 276 \times 0.01 \times 1 = 2.76$

$\log T\% = 2 - A = -0.76$

$T\% = 0.17\%$

6. The $1s^2$ electrons form a closed shell and need not be considered.

2s : $l = 0$, $s = \frac{1}{2}$; therefore, $j = \frac{1}{2}$; $^2S_{1/2}$

2p : $l = 1$, $s = \frac{1}{2}$; therefore, $j = \frac{3}{2}$ and $\frac{1}{2}$;

$^2P_{3/2}$, $^2P_{1/2}$

7. The closed shells need not be considered.

$l_1 = l_2$ so that $L = 0, 1, 2$.

$s_1 = s_2 = \frac{1}{2}$ so that $S = 0, 1$.

When $S = 0$ we therefore have ^1S, ^1P and ^1D

When $S = 1$ we have ^3S, ^3P, ^3D.

For the singlet terms $J = L$ and therefore only $^1S_0$, $^1P_1$ and $^1D_2$ occur.

For ^3D, since $L = 2$ and $S = 1$ we have

$J = 3, 2, 1$; therefore $^3D_3$, $^3D_2$, $^3D_1$

For ^3P, $J = 2, 1, 0$ and therefore $^3P_2$, $^3P_1$, $^3P_0$

For ^3S, $J = 1$ only and therefore $^3S_1$.

8. (a) For the 3p electron $l = 1$, and $s = \frac{1}{2}$; we therefore have the term ^2P. The j values are $1 + \frac{1}{2} = \frac{3}{2}$ and $1 - \frac{1}{2} = \frac{1}{2}$; therefore the terms are

$$^2P_{3/2} \text{ and } ^2P_{1/2}$$

(b) For $3d^1$, $l = 2$ and $s = \frac{1}{2}$; therefore the terms are

$$^2D_{5/2},\ ^2D_{3/2} \text{ and } ^2D_{1/2}$$

9. ^1P: $S = 0,\ L = 1;\ J = 1$

^3P: $S = 1,\ L = 1;\ J = 2, 1, 0$

^4P: $S = \frac{3}{2},\ L = 1;\ J = \frac{5}{2}, \frac{3}{2}, \frac{1}{2}$

^1D: $S = 0,\ L = 2;\ J = 2$

^2D: $S = \frac{1}{2},\ L = 2;\ J = \frac{5}{2}, \frac{3}{2}$

^3D: $S = 1,\ L = 2;\ J = 3, 2, 1$

^4D: $S = \frac{3}{2},\ L = 2;\ J = \frac{7}{2}, \frac{5}{2}, \frac{3}{2}, \frac{1}{2}$

10. The Lande g factor is

$$g_{1/2} = 1 + \frac{\left(\frac{1}{2} \times \frac{3}{2}\right) + \left(\frac{1}{2} \times \frac{3}{2}\right) - (1 \times 2)}{2\left(\frac{1}{2} \times \frac{3}{2}\right)} = \frac{2}{3}$$

The $^2P_{1/2}$ level splits into two levels with $M_J = \frac{1}{2}$ $M_J = -\frac{1}{2}$. The energy is given by Eq. 13.48:

$$E = g_J \mu_B B M_J$$

and the separation is therefore

$$\Delta E = g_J \mu_B B$$

$$= \frac{2}{3} \times 9.273 \times 10^{-24} \times 4.0 \text{ J}$$

$$= 2.47 \times 10^{-23} \text{ J}$$

To convert to cm^{-1} we divide by hc with $c = 2.998 \times 10^{10} \text{ cm s}^{-1}$:

$$\Delta E = \frac{2.47 \times 10^{-23} \text{ J}}{6.626 \times 10^{-34} \text{ J s} \times 2.998 \times 10^{10} \text{ cm s}^{-1}}$$

$$= 1.24 \text{ cm}^{-1}$$

11. For the $^3P_0$ level, $g_J = 0$. The splitting of the line is therefore entirely due to the splitting of the $^3D_1$ level. For this level

$$g_J = 1 + \frac{(1 \times 2) + (1 \times 2) - (2 \times 3)}{(2 \times 2)} = \frac{1}{2}$$

It will be split into three levels with $M_J = 1, 0, -1$, and the separation between the levels is

$$\Delta E = g_J \mu_B B$$

$$= \frac{1}{2} \times 9.273 \times 10^{-24} \times 4.0 = 1.85 \times 10^{-23} \text{ J}$$

$$= \frac{1.85 \times 10^{-23} \text{ J}}{6.626 \times 10^{-34} \text{ J} \times 2.998 \times 10^{10} \text{ cm s}^{-1}}$$

$$= 0.93 \text{ cm}^{-1}$$

12. The separation is $2B$ and therefore

$$B = 0.5115 \text{ cm}^{-1}$$

This is equal to $h/8\pi^2 Ic$, and the moment of inertia is therefore

$$I = \frac{6.626 \times 10^{-34} \text{ J s}}{0.5115 \text{ cm}^{-1}(8\pi^2)(2.998 \times 10^{10} \text{ cm s}^{-1})}$$
$$= 5.472 \times 10^{-46} \text{ kg m}^2$$

The reduced mass (Eq. 9.73) is

$$\mu = \frac{35 \times 19 \text{ g mol}^{-1}}{54 \times 6.022 \times 10^{23} \text{ mol}^{-1}} = 2.045 \times 10^{-26} \text{ kg}$$

The moment of inertia is μr^2 and therefore

$$r = \sqrt{\frac{5.472 \times 10^{-46}}{2.045 \times 10^{-26}}} = 1.64 \times 10^{-10} \text{ m}$$
$$= 164 \text{ pm}$$

13. $B = 20.85 \text{ cm}^{-1}$ and the moment of inertia is

$$I = \frac{6.626 \times 10^{-34}}{20.85 \times 8\pi^2 \times 2.998 \times 10^{10}} = 1.343 \times 10^{-47} \text{ kg m}^2$$

The reduced mass is

$$\mu = \frac{1.008 \times 19.0 \text{ g}}{20.008 \times 6.022 \times 10^{23}} = 1.590 \times 10^{-23} \text{ kg}$$

The interatomic distance is thus

$$r = \sqrt{\frac{1.343 \times 10^{-47}}{1.590 \times 10^{-27}}} = 9.19 \times 10^{-11} \text{ m}$$

$$= 92 \text{ pm}$$

The separation is inversely proportional to I and therefore to the reduced mass; the interatomic separations are assumed to be the same. The reduced masses for HF, DF and TF are in the ratio

$$\frac{1 \times 19}{20} : \frac{2 \times 19}{21} : \frac{3 \times 19}{22}$$

$$= \quad 1 \quad : \quad 1.90 \quad : \quad 2.72$$

The predicted separations are therefore

DF : $22.1\ \text{cm}^{-1}$

TF : $15.4\ \text{cm}^{-1}$

14. The separation is $2B$; therefore

$$B = 5.7635 \times 10^{10}\ \text{s}^{-1}$$

The moment of inertia is therefore

$$I = \frac{6.626 \times 10^{-34}\ \text{J s}}{8\pi^2 \times 5.7635 \times 10^{10}\ \text{s}^{-1}} = 1.456 \times 10^{-46}\ \text{kg m}^2$$

The reduced mass is

$$\mu = \frac{12.000 \times 15.995\ \text{g}}{27.995 \times 6.022 \times 10^{23}} = 1.1385 \times 10^{-26}\ \text{kg}$$

The interatomic distance is

$$r = \sqrt{\frac{1.456 \times 10^{-46}}{1.1385 \times 10^{-26}}} = 1.131 \times 10^{-10}$$

$$= 113\ \text{pm}$$

15.

	Pure rotational spectrum	Vibrational-rotational spectrum	Rotational Raman spectrum	Vibrational Raman spectrum
H_2			X	X
HCl	X	X	X	X
CO_2		X	X	X
CH_4		X		X
H_2O	X	X	X	X
CH_3Cl	X	X	X	X
CH_2Cl_2	X	X	X	X
H_2O_2	X	X	X	X
NH_3	X	X	X	X
SF_6		X		X

16. The force constant is related to the fundamental frequency by Eq. 13.80:

$$k = 4\pi^2 \nu_0^2 \mu$$

The reduced mass (see example on p. 588 of the text) is

$$\mu = 1.614 \times 10^{-27} \text{ kg}$$

The frequency ν is

$$2988.9 \text{ cm}^{-1} \times 2.998 \times 10^{10} \text{ cm s}^{-1}$$

$$= 8.961 \times 10^{13} \text{ s}^{-1}$$

The force constant is thus

$$k = 4\pi^2 (8.961 \times 10^{13} \text{ s}^{-1})^2 \, 1.614 \times 10^{-27} \text{ kg}$$

$$= 511.6 \text{ kg s}^{-2}$$

17. The separation is $4B$ (Figure 13.21) and B is therefore $0.2438\ \text{cm}^{-1}$. The moment of inertia is therefore

$$I = \frac{6.626 \times 10^{-34}\ \text{J s}}{0.2438\ \text{cm}^{-1}\ (8\pi^2)\ 2.998 \times 10^{10}\ \text{cm s}^{-1}}$$

$$= 1.148 \times 10^{-45}\ \text{kg m}^2$$

The reduced mass is

$$\mu = \frac{35 \times 35 \times 10^{-3}}{70 \times 6.022 \times 10^{23}} = 2.906 \times 10^{-26}\ \text{kg}$$

The interatomic distance is thus

$$r = \sqrt{\frac{1.148 \times 10^{-45}}{2.906 \times 10^{-26}}} = 1.99 \times 10^{-10}\ \text{m} = 199\ \text{pm}$$

18. The zero-point energy of H_2 is

$$\frac{1}{2}h\nu_0 = \frac{1}{2}(6.626 \times 10^{-34}\ \text{J s})(1.257 \times 10^{14}\ \text{s}^{-1})$$

$$= 4.164 \times 10^{-20}\ \text{J} = 25.1\ \text{kJ mol}^{-1}$$

Classical dissociation energy $= 432.0 + 25.1$

$$= 457.1\ \text{kJ mol}^{-1}$$

Reduced masses of H_2, HD and D_2 are in the ratio

$$\frac{1}{2} : \frac{2}{3} : 1$$

$$= 1 : \frac{4}{3} : 2$$

Frequencies are inversely proportional to $\sqrt{\mu}$:

estimated $\nu_0(\text{HD}) = 1.089 \times 10^{14}\ \text{s}^{-1}$

$$\frac{1}{2}h\nu_0(\text{HD}) = 21.7 \text{ kJ mol}^{-1}$$

estimated $\nu_0(D_2) = 8.89 \times 10^{13}\ s^{-1}$

$$\frac{1}{2}h\nu_0(D_2) = 17.7 \text{ kJ mol}^{-1}$$

Estimated dissocation energies:

HD: $457.1 - 21.7 = 435.4$ kJ mol^{-1}

D_2: $457.1 - 17.7 = 439.4$ kJ mol^{-1}

19. The symmetric molecule has only two modes that are active in the infrared:

```
B →← A —— B →
```

and

```
          ↑
B —— A —— B    (two degenerate modes)
↓              ↓
```

The symmetric stretch is inactive.

The unsymmetrical molecules have three modes active in the infrared:

```
← A —— B —— B →

A →← B —— B

          ↑
B —— A —— B    (two degenerate modes)
↓              ↓
```

The molecule is therefore A-B-B.

20. The reduced masses, $m_1m_2/(m_1 + m_2)$, are

O-H: $\mu_{OH} = \dfrac{1 \times 16}{(1 + 16)\ L}$ g

O-D: $\mu_{OD} = \dfrac{2 \times 16}{(2 + 16)\ L}$ g

and are in the ratio

$$\frac{\mu_{OD}}{\mu_{OH}} = \frac{2 \times 17}{18} = 1.89$$

The force constants are the same, and the frequencies are inversely proportional to the square root of the reduced mass. Thus

$$\bar{\nu}_{OD} = \bar{\nu}_{OH}/(1.89)^{1/2} = \frac{3300}{1.37}\ \text{cm}^{-1} = 2400\ \text{cm}^{-1}$$

21. The wave numbers corresponding to the two wavelengths are

435.83 nm: $\dfrac{1}{435.83 \times 10^{-9}}\ \text{m}^{-1} = 22\ 944.7\ \text{cm}^{-1}$

476.85 nm: $\dfrac{1}{476.85 \times 10^{-9}}\ \text{m}^{-1} = 20\ 971.0\ \text{cm}^{-1}$

The difference, 1973.7 cm^{-1}, corresponds to a vibration in the C_2H_2 molecule.

22. The reduced mass of $H^{127}I$ is

$$\mu = \frac{1.008 \times 126.9}{127.908 \times 6.022 \times 10^{26}}\ \text{kg}$$

$$= 1.6607 \times 10^{-27}\ \text{kg}$$

The frequency ν is

$$2309.5\ \text{cm}^{-1} \times 2.998 \times 10^{10}\ \text{cm s}^{-1}$$

$$= 6.924 \times 10^{13}\ \text{s}^{-1}$$

The force constant k is

$$k = 4\pi^2\nu^2\mu \qquad \text{(Eq. 13.77)}$$

$$= 4\pi^2(6.924 \times 10^{13}\ \text{s}^{-1})^2 \times 1.6607 \times 10^{-27}\ \text{kg}$$

$= 314.3 \text{ kg s}^{-1}$

23. (a) The curves are similar in both states:

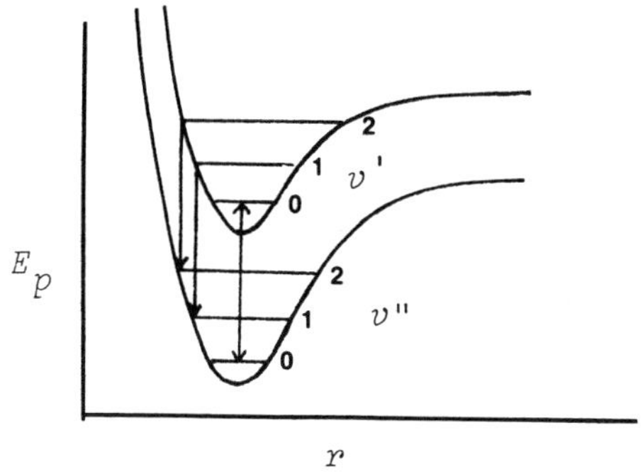

(b)

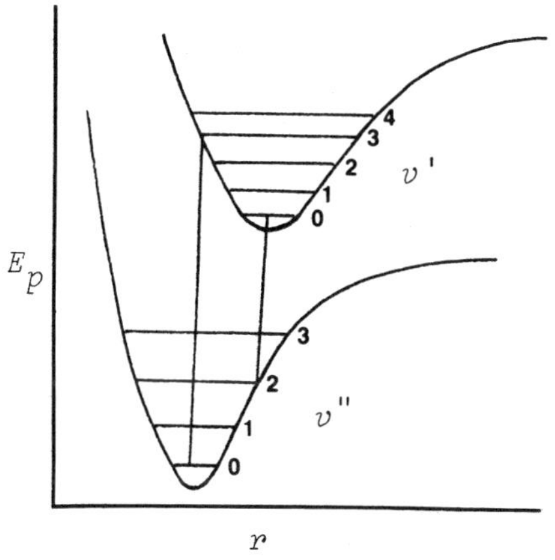

(c) Absorption from $v'' = 0$ gives the upper state with enough energy to dissociate at once. Emission occurs from lower vibrational states.

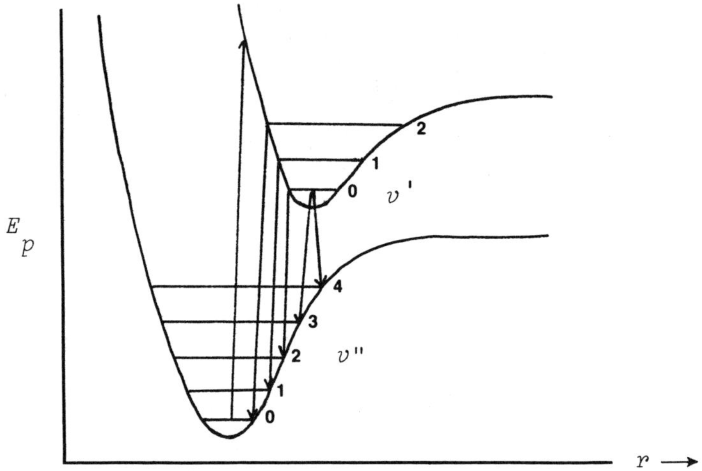

(d) Predissociation

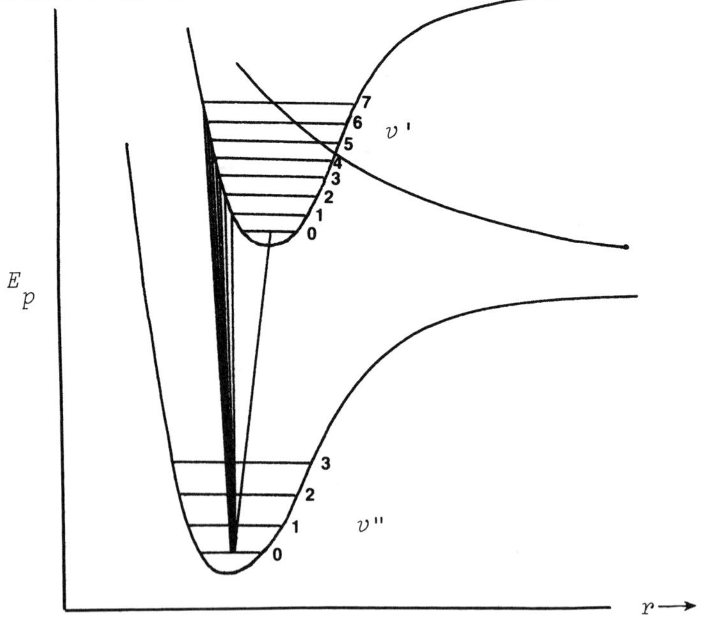

24. From Eq. 13.145 it follows that

$$T = \left(\frac{\Delta\lambda\ c}{2\lambda}\right)^2 \cdot \frac{m}{2k}$$

The molecular mass m is

$$\frac{56.94 \times 10^{-3}}{6.022 \times 10^{23}} \text{ kg} = 9.46 \times 10^{-26} \text{ kg}$$

Therefore

$$T = \frac{(0.053 \text{ nm } 2.998 \times 10^{8} \text{ m s}^{-1})(9.46 \times 10^{-26} \text{ kg})}{(2 \times 677.4 \text{ nm})^2 \ 2 \times 1.381 \times 10^{-23} \text{ J K}^{-1}}$$

$$= 4.7 \times 10^{5} \text{ K}$$

25. From Eq. 13.147

$$\Delta\tau/\text{s} = \frac{2.7 \times 10^{-12}}{\Delta\bar{\nu}/\text{cm}^{-1}}$$

(a) $\Delta\tau = 2.7 \times 10^{-10}$ s

(b) $\Delta\tau = 2.7 \times 10^{-11}$ s

(c) $\Delta\tau = 2.7 \times 10^{-12}$ s

(d) $$\Delta\tau = \frac{2.7 \times 10^{-12} \ 2.998 \times 10^{10} \text{ cm s}^{-1}}{200 \times 10^{6} \text{ s}^{-1}}$$

$$= 4.05 \times 10^{-10} \text{ s}$$

26.

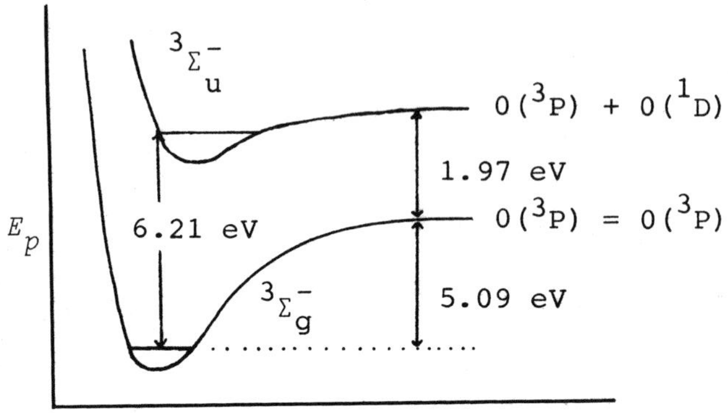

Dissociation energy of O_2 ($^3\Sigma_u^-$) into $O(^3P) + O(^1D)$

$= 5.09 + 1.97 - 6.21 = 0.85$ eV $= 82.0$ kJ mol^{-1}

27. The wavenumber of 20 302.6 cm^{-1} corresponds to a frequency of

$$\nu = 20\ 302.6\ \text{cm}^{-1} \times 2.998 \times 10^{10}\ \text{cm s}^{-1}$$

$$= 6.087 \times 10^{14}\ \text{s}^{-1}$$

and to an energy of

$$h\nu = 4.033 \times 10^{-19}\ \text{J} = 2.518\ \text{eV}$$

The wavelength 589.3 nm corresponds to a frequency of

$$\nu = \frac{2.998 \times 10^{10}\ \text{m s}^{-1}}{589.3 \times 10^{-9}\ \text{m}} = 5.087 \times 10^{14}\ \text{s}^{-1}$$

and to an energy of

$$h\nu = 3.3709 \times 10^{-19}\ \text{J} = 2.104\ \text{eV}$$

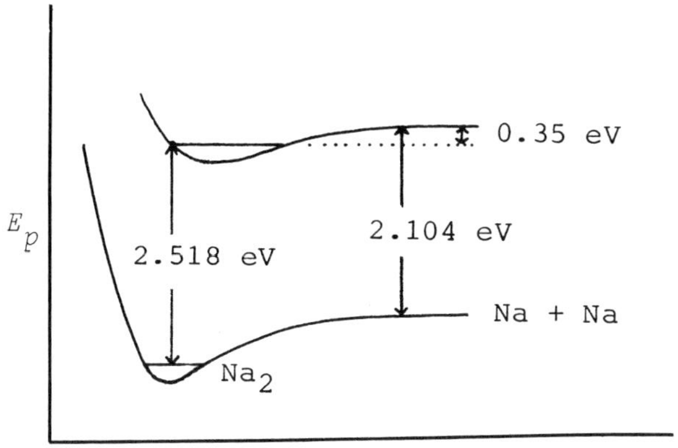

Dissociation energy of Na_2 = 2.518 + 0.35 - 2.104 eV

$$= 0.76 \text{ eV} = 73.7 \text{ kJ mol}^{-1}$$

28. A wavelength of 8.00 mm corresponds to

$$\nu = \frac{2.998 \times 10^8 \text{ m s}^{-1}}{0.008 \text{ m}} = 3.748 \times 10^{10} \text{ s}^{-1}$$

From Eq. 13.151,

$$B = \frac{h\nu}{g\mu_B}$$

$$= \frac{6.626 \times 10^{-34} \text{ J s} \times 3.748 \times 10^{10} \text{ s}^{-1}}{2.023 \times 9.274 \times 10^{-24} \text{ J T}^{-1}}$$

$$= 1.324 \text{ T}$$

29. From Eq. 13.151

$$g = \frac{h\nu}{\mu_B B}$$

$$= \frac{6.626 \times 10^{-34} \text{ J s} \times 10.42 \times 10^9 \text{ s}^{-1}}{9.274 \times 10^{-24} \text{ J T}^{-1} \times 0.3715 \text{ T}}$$

$$= 2.0026$$

30.

	^2H	^{19}F	^{35}Cl	^{37}Cl
Spin, I	1	$\frac{1}{2}$	$\frac{3}{2}$	$\frac{3}{2}$
M_I values	1,0,-1	$+\frac{1}{2},-\frac{1}{2}$	$+\frac{3}{2},+\frac{1}{2},-\frac{1}{2},-\frac{3}{2}$	$+\frac{3}{2},+\frac{1}{2},-\frac{1}{2},-\frac{3}{2}$
Number of lines	3	2	4	4
g_N	0.857	5.257	0.550	0.456
$\mu_N/10^{-27}$ J T^{-1}	4.329	13.28	4.170	3.455

31. The frequency of 60×10^6 s^{-1} corresponds to an energy of

$6.626 \times 10^{-34} \times 60 \times 10^6$ J $= 3.976 \times 10^{-26}$ J

Resonance could be observed with the proton and with ^{35}Cl. For the proton

$\Delta E = g_N\mu_N B = 5.586 \times 5.0508 \times 10^{-27} \times B$ (J T^{-1})

Resonance would thus be observed at

$$B = \frac{3.976 \times 10^{-26}}{5.586 \times 5.0508 \times 10^{-27}} \text{ T}$$

$= 1.41$ T

For ^{35}Cl,

$\Delta E = 0.5479 \times 5.0508 \times 10^{-27}$ B

Resonance would thus be found at

$$B = \frac{3.976 \times 10^{-26}}{0.5479 \times 5.0508 \times 10^{-27}} \text{ T}$$

$= 14.4$ T

(This is outside the usual instrumental range)

32. From Eq. 13.162,

$$B_{eff} = (1 - \sigma)B$$

Thus

$$\Delta B_{eff} = -\Delta\sigma B$$

$$= -(9.80 - 2.20) \times 10^{-6} \times 1.5 \text{ T}$$

$$= -1.14 \times 10^{-5} \text{ T} = -11.4 \ \mu\text{T}$$

$$\Delta\sigma = (9.80 - 2.20) \times 10^{-6} = 7.60 \times 10^{-6}$$

and this gives a frequency splitting of

$$7.60 \times 10^{-6} \times 60 \text{ MHz} = 456 \text{ Hz}$$

33. There are four orientations, corresponding to

$$M_I = +\frac{3}{2}, +\frac{1}{2}, -\frac{1}{2}, -\frac{3}{2}$$

The resonance corresponds to $\Delta M_I = 1$ and is at an energy given by Eq. 13.157;

$$\Delta E = g_N \mu_N B$$

$$= 0.2606 \times 5.0508 \times 10^{-27} \times 1.0 \text{ J} = 1.316 \times 10^{-27} \text{ J}$$

This corresponds to a frequency of

$$\nu = \frac{1.316 \times 10^{-27}}{6.626 \times 10^{-34}} = 1.986 \times 10^{6} \text{ s}^{-1} = 1.99 \text{ MHz}$$

34. The frequency 60 MHz corresponds to an energy of

$$60 \times 10^{6} \times 6.626 \times 10^{-34} = 3.976 \times 10^{-26} \text{ J}$$

Resonance is thus found at a field of

$$B = \frac{3.976 \times 10^{-26}}{1.792 \times 5.0508 \times 10^{-27}} \text{ T}$$

$$= 4.39 \text{ T}$$

35. From Eq. 13.146

$$\Delta E = \frac{h}{4\pi\Delta\tau} = \frac{6.626 \times 10^{-34} \text{ J s}}{4\pi \times 2 \times 10^{-7} \text{ s}}$$

$$= 2.636 \times 10^{-28} \text{ J}$$

The uncertainty in the frequency is thus

$$\Delta\nu = \frac{2.636 \times 10^{-28} \text{ J}}{6.626 \times 10^{-34} \text{ J s}} = 3.98 \times 10^{5} \text{ Hz}$$

The uncertainty in the wavenumber is

$$\Delta\bar{\nu} = \frac{3.98 \times 10^{5} \text{ s}^{-1}}{2.998 \times 10^{10} \text{ cm s}^{-1}} = 1.33 \times 10^{-5} \text{ cm}^{-1}$$

36. *cis*-$C_2H_2Cl_2$: The point group is C_{2v}:

(a) a_1; active in both infrared and Raman

(b) a_2; inactive in infrared but active in Raman

(c) b_1; active in infrared and Raman

(d) a_1; active in both infrared and Raman

trans-$C_2H_2Cl_2$: the point group is C_{2h}:

(e) a_g; inactive in infrared, active in Raman

(f) a_u; active in infrared, inactive in Raman

(g) b_u; active in infrared, inactive in Raman

(h) a_g; inactive in infrared, active in Raman

Benzene: The point group is D_{6h}:

(i) a_{1g}; inactive in infrared, active in Raman

(j) a_{2u}; active in infrared, inactive in Raman

(k) b_{1u}; inactive in both infrared and Raman

37. The B values are half the separations:

$$B(^{16}O^{12}C^{32}S) = 6.0815 \times 10^9 \ s^{-1}$$

$$B(^{16}O^{12}C^{34}S) = 5.9325 \times 10^9 \ s^{-1}$$

The moments of inertia are

$$I(^{16}O^{12}C^{32}S) = \frac{6.626 \times 10^{-34} \ J \ s}{8\pi^2 \times 6.0815 \times 10^9 \ s^{-1}}$$

$$= 1.3799 \times 10^{-45} \ kg \ m^2$$

$$I(^{16}O^{12}C^{34}S) = \frac{6.626 \times 10^{-34} \ J \ s}{8\pi^2 \times 5.9325 \times 10^9 \ s^{-1}}$$

$$= 1.4146 \times 10^{-45} \ kg \ m^2$$

The moment of inertia of a linear triatomic molecule is given by Eq. 9.62 and we write x for r_{12} and y for r_{23}, and M_1, M_2 and M_3 for the molar masses

$$\frac{I(^{16}O^{12}C^{32}S)}{L} = M_1x^2 + M_3y^2 - \frac{(M_1x - M_3y)^2}{M}$$

$$= 1.3799 \times 10^{-42} \ g \ m^2$$

$$\frac{I(^{16}O^{12}C^{34}S)}{L} = M_1x^2 + M_3'y^2 - \frac{(M_1x - M_3y)^2}{M}$$

$$= 1.4146 \times 10^{-42} \ g \ m^2$$

Insertion of $L = 6.022 \times 10^{23} \ mol^{-1}$, $M_1 = 16 \ g \ mol^{-1}$, $M_3 = 32 \ g \ mol^{-1}$, $M_3' = 34 \ g \ mol^{-1}$, $M = 60 \ g \ mol^{-1}$, $x = 116$ pm and $y = 156$ pm, into the left-hand sides of these equations, gives approximately the values on the right-hand side. For example, for the first equation

$$\text{L.H.S.} = 16(116)^2 + 32(156)^2 - \frac{[(16 \times 116) - (32 \times 156)]^2}{60}$$

$$= 830\ 140 \text{ g pm}^2 \text{ mol}^{-1} = 1.3785 \times 10^{-42} \text{ g m}^2$$

38. The planar configuration must arise from sp^2 hybridization; the point group is D_{3h}.
Since planar CH_3 has no dipole moment it shows no microwave spectrum.
The vibrations are as shown in Figure 13.19a for BF_3. As discussed in the text (p. 603), all normal modes except the completely symmetric one (a_1') give an infrared spectrum; i.e. an infrared spectrum is given by the a_2'' and e' vibrations.

39. At small x values the exponential in Eq. 13.101 may be expanded to $1 - \beta x$ and the potential energy is thus given by

$$E_p = D_e \beta^2 x^2$$

Then

$$\frac{dU}{dx} = 2D_e \beta^2 x$$

The restoring force f is $-dE_p/dx$:

$$f = -2D_e \beta^2 x = -kx$$

where k is the force constant. Thus

$$k = 2D_e \beta^2$$

The frequency ν_0 is

$$\nu_0 = \frac{1}{2\pi}\sqrt{\frac{2D_e\beta^2}{\mu}}$$

The reduced mass μ was calculated for $H^{35}Cl$ in the example on p. 588:

$$\mu = 1.614 \times 10^{-27}\ \text{kg}$$

Then

$$\nu_0 = \frac{1}{2\pi}\sqrt{\frac{2 \times 4.67 \times 1.602 \times 10^{-19}\ \text{J}(1.85 \times 10^{10}\ \text{m}^{-1})^2}{1.614 \times 10^{-27}\ \text{kg}}}$$

$$= 8.96 \times 10^{13}\ \text{s}^{-1}$$

CHAPTER 14: WORKED SOLUTIONS
MOLECULAR STATISTICS

1. (a) $$\frac{\sqrt{\overline{u^2}}}{\overline{u}} = \sqrt{\frac{3\pi}{8}} = 1.085$$

(b) $$\frac{\overline{u}}{u_{mp}} \quad \frac{2}{\sqrt{\pi}} = 1.128$$

The differences between $\sqrt{\overline{u^2}}$ and $\overline{u}$, and between $\overline{u}$ and u_{mp} increase with T and decrease with m.

2. From Table 14.2 the average speed is

$$\overline{u} = \sqrt{\frac{8\mathbf{k}T}{\pi m}}$$

and therefore

$$T = \frac{\pi \; m \; \overline{u}^2}{8\mathbf{k}}$$

(a) The mass of the hydrogen molecule is

$$\frac{2.016 \times 10^{-3}}{6.022 \times 10^{23}} = 3.347 \times 10^{-27} \text{ kg}$$

$$T = \frac{\pi \times 3.347 \times 10^{-27} \text{ kg} \times (1.07 \times 10^4 \text{ m s}^{-1})^2}{8 \times 1.381 \times 10^{-23} \text{ J K}^{-1}}$$

$$= 10\ 900 \text{ K}$$

(b) $$m_{O_2} = \frac{32.00 \times 10^{-3}}{6.022 \times 10^{23}} = 5.31 \times 10^{-26} \text{ kg}$$

$$T = \frac{\pi \times 5.31 \times 10^{-26} \times (1.07 \times 10^4)^2}{8 \times 1.381 \times 10^{-23}} = 173\ 000 \text{ K}$$

3. (a) From Eq. 14.47, the ratio for two different speeds u_1 and u_2 is

$$\frac{e^{-mu_1^2/2\mathbf{k}T} \; u_1^2}{e^{-mu_2^2/2\mathbf{k}T} \; u_2^2} = \left(\frac{u_1}{u_2}\right)^2 e^{-m(u_1^2 - u_2^2)/2\mathbf{k}T}$$

$$u_1 = 2\bar{u} = 2\sqrt{\frac{8\mathbf{k}T}{\pi m}};\quad u_2 = \bar{u} = \sqrt{\frac{8\mathbf{k}T}{\pi m}}$$

$$\left(\frac{u_1}{u_2}\right)^2 = 4 \text{ and } u_1^2 - u_2^2 = \frac{3 \times 8\mathbf{k}T}{\pi m}$$

$$\frac{m(u_1^2 - u_2^2)}{2\mathbf{k}T} = \frac{12}{\pi}$$

The ratio is $4\ e^{-12/\pi} = 8.77 \times 10^{-2}$

There is no effect of mass or temperature

(b) For two temperatures T_1 and T_2 the ratio is, from Eq. 14.47

$$\left(\frac{T_2}{T_1}\right)^{3/2} \frac{e^{-m\bar{u}_1^2/\mathbf{k}T_1}}{e^{-m\bar{u}_2^2/2\mathbf{k}T_2}} \left(\frac{u_1}{u_2}\right)^2$$

Insertion of the expression for $\bar{u}_1$, and $\bar{u}_2$ gives

$$\left(\frac{T_2}{T_1}\right)^{3/2} \frac{e^{-4/\pi}}{e^{-4/\pi}} \left(\frac{T_1}{T_2}\right) = \left(\frac{T_2}{T_1}\right)^{1/2}$$

With T_1 = 100°C = 373.15 K and T_2 = 298.15 K this gives

$$(298.15/373.15)^{1/2} = 0.89$$

There is no dependence on mass.

4. The average molecular speed is

$$\bar{u} = \left(\frac{8\mathbf{k}T}{\pi m}\right)^{1/2}$$

The ideal gas law is

$$PV = nRT = nL\mathbf{k}T$$

and the density ρ is

$$\rho = \frac{nmL}{V}$$

Thus

$$\bar{u} = \left(\frac{8PV}{nL\pi} \cdot \frac{nL}{\rho V}\right)^{1/2} = \left(\frac{8P}{\pi\rho}\right)^{1/2}$$

Since P and ρ are the same in the two gases the average speeds are the same.

5. (a) Call 25°C T_2 and 100°C T_1. Then the average speed $\bar{u}_{25°}$ is

$$\bar{u}_{25°} = \sqrt{\frac{8\mathbf{k}T_2}{\pi m}}$$

From Eq. 14.47 the required ratio is

$$\left(\frac{T_2}{T_1}\right)^{3/2} \frac{e^{-8\mathbf{k}T_2 m/2\mathbf{k}T_1 \pi m}}{e^{-8\mathbf{k}m/2\mathbf{k}\pi m}}$$

$$= \left(\frac{T_2}{T_1}\right)^{3/2} e^{-\frac{4}{\pi}\left(\frac{T_2}{T_1} - 1\right)}$$

$$= \left(\frac{298.15}{373.15}\right)^{3/2} e^{0.2559} = 0.922$$

(b) The ratio for a speed of 10 $\bar{u}_{25°}$ is

$$\left(\frac{T_2}{T_1}\right)^{1/2} e^{-\frac{400}{\pi}\left(\frac{T_2}{T_1} - 1\right)}$$

$$= \left(\frac{298.15}{373.15}\right)^{3/2} e^{25.59} = 9.29 \times 10^{10}$$

6. $C_P = \left(\frac{\partial H}{\partial T}\right)_P$

$$= \left[\frac{\partial}{\partial T}\left\{\mathbf{k}T^2\left(\frac{\partial \ln Q}{\partial T}\right)_V\right\}\right]_P + N\mathbf{k}$$

7. (a) $Q = q^N; \ \ln Q = N \ln q$

$$U - U_o = N\mathbf{k}T^2\left(\frac{\partial \ln q}{\partial T}\right)_V$$

$$S = N\mathbf{k}T\left(\frac{\partial \ln q}{\partial T}\right)_V + N\mathbf{k} \ln q$$

$$A - U_o = -N\mathbf{k}T \ln q$$

$$H - U_o = N\mathbf{k}T^2\left(\frac{\partial \ln q}{\partial T}\right)_V + N\mathbf{k}T$$

$$G - U_o = -N\mathbf{k}T \ln q + N\mathbf{k}T = -N\mathbf{k}T \ln(q/\mathrm{e})$$

(b) $Q = q^N/N!; \ \ln Q = \ln q^N - \ln N!$

$$= N \ln q - N \ln N + N$$

$$U - U_o = N\mathbf{k}T^2\left(\frac{\partial \ln q}{\partial T}\right)_V$$

$$S = N\mathbf{k}T\left(\frac{\partial \ln q}{\partial T}\right)_V + N\mathbf{k} \ln q - N\mathbf{k} \ln N + N\mathbf{k}$$

$$= N\mathbf{k}T(\partial \ \ln q/\partial T)_V + N\mathbf{k} \ln(q\mathrm{e}/N)$$

$$A - U_o = -N\mathbf{k}T \ln q + N\mathbf{k}T \ln N - N\mathbf{k}T$$

$$= -N\mathbf{k}T \ln(q\mathrm{e}/N)$$

$$H - U_o = N\mathbf{k}T^2\left(\frac{\partial \ln q}{\partial T}\right)_V + N\mathbf{k}T$$

$$G - U_o = -N\mathbf{k}T \ln q + N\mathbf{k}T \ln N - N\mathbf{k}T + N\mathbf{k}T$$

$$= -N\mathbf{k}T \ln(q/N)$$

8. From Eq. 3.118

$$P = -\left(\frac{\partial A}{\partial V}\right)_T = \mathbf{k}T\left(\frac{\partial \ln Q}{\partial V}\right)_T$$

(a) $\ln Q = N \ln q$

$$P = N\mathbf{k}T\left(\frac{\partial \ln q}{\partial V}\right)_T$$

$$= nRT\left(\frac{\partial \ln q}{\partial V}\right)_T$$

(b) $\ln q = N \ln q - N \ln N + N$

This gives the same expression for P as above.

9. $$S = \mathbf{k}T\left(\frac{\partial \ln Q}{\partial T}\right)_V + \mathbf{k} \ln Q$$

For distinguishable molecules $\ln Q = N \ln q$

and $$S_{dist} = N\mathbf{k}T\left(\frac{\partial \ln q}{\partial T}\right)_V + N\mathbf{k} \ln q$$

For indistinguishable molecules $\ln Q = N \ln q - N \ln N + N$

and $$S_{indis} = N\mathbf{k}T\left(\frac{\partial \ln q}{\partial T}\right)_V + N\mathbf{k} \ln q - N\mathbf{k} \ln N + N\mathbf{k}$$

Thus $S_{dist} - S_{indis} = N\mathbf{k} \ln N + N\mathbf{k}$

For 1 mol, $N = 6.022 \times 10^{23}$ and $\mathbf{k} = 1.381 \times 10^{-23}$ J K^{-1}

Then $(S_{dist} - S_{indis})_m = 8.316 \ln(6.022 \times 10^{23}) + 8.316$

$$= 463.7 \text{ J K}^{-1} \text{ mol}^{-1}$$

10. The system partition function for a two-dimensional

monatomic gas is

$$Q = \frac{2\pi m\mathbf{k}T}{h^2}$$

Then from Eq. 14.67, for 1 mol,

$$U_m - U_{0,m} = RT^2\left(\frac{\partial \ln Q}{\partial T}\right)_V = RT$$

If T = 25.0°C = 298.15 K,

$$U_m - U_{0,m} = 8.314 \times 298.15 \text{ J mol}^{-1} = 2480 \text{ J mol}$$

$$= 2.48 \text{ kJ mol}^{-1}$$

11. In the limit as T → ∞ the vibrational partition function becomes

$$\frac{1}{1 - (1 - h\nu/\mathbf{k}T)} = \frac{\mathbf{k}T}{h\nu}$$

Then, as in the preceding problem

$$U_m - U_{0,m} = RT$$

and

$$C_{V,m} = R = 8.314 \text{ J K}^{-1} \text{ mol}^{-1}$$

12. $q = \frac{(2\pi m\mathbf{k}T)^{3/2}V}{h^3}$

For N indistinguishable molecules

$$\ln Q = N \ln q - N \ln N + N$$

$$U - U_0 = \mathbf{k}T^2\left(\frac{\partial \ln Q}{\partial T}\right)_V = N\mathbf{k}T^2\left(\frac{\partial \ln q}{\partial T}\right)_V$$

$$= N\mathbf{k}T^2 \cdot \frac{3}{2} \cdot \frac{1}{T} = \frac{3}{2} N\mathbf{k}T$$

For 1 mol, $U_m - U_{0,m} = \frac{3}{2} RT$

13. The molecular partition function with $V = 1\ m^3$ and $T = 300$ K is

$$q_t = \frac{(2\pi \times 1.381 \times 10^{-23} \times 300)^{3/2}}{(6.625 \times 10^{-34})^3} (m/\text{kg})^{3/2}$$

$$= 1.444 \times 10^{70}\ (m/\text{kg})^{3/2}$$

(a) For N_2, $m = 2 \times 14.0067 \times 10^{-3}/6.022 \times 10^{23}$

$= 4.652 \times 10^{-26}$ kg

$q_t = 1.449 \times 10^{32}$

(b) For H_2O, $m = 18.015 \times 10^{-3}/6.022 \times 10^{23}$

$= 2.992 \times 10^{-26}$ kg

$q_t = 7.47 \times 10^{31}$

(c) For C_6H_6, $m = 78.114 \times 10^{-3}/6.022 \times 10^{23}$

$= 1.297 \times 10^{-25}$ kg

$q_t = 6.746 \times 10^{32}$

$\ln Q = N \ln q_t - N \ln N + N$

For the molar translational partition function $Q_{t,m}$:

(a) $\ln Q_{t,m} = 4.4595 \times 10^{25} - 3.297 \times 10^{25} + 6.022 \times 10^{23} = 1.223 \times 10^{25}$

(b) $\ln Q_{t,m} = 4.420 \times 10^{25} - 3.297 \times 10^{25} + 6.022 \times 10^{23} = 1.183 \times 10^{25}$

(c) $\ln Q_{t,m} = 4.552 \times 10^{25} - 3.297 \times 10^{25} + 6.022 \times 10^{23} = 1.315 \times 10^{25}$

14. Mass of N atom $= 14.0067 \times 10^{-3}/6.022 \times 10^{23}$

$$= 2.326 \times 10^{-26} \text{ kg}$$

Moment of inertia of N_2

$$I = \frac{1}{2} \times 2.326 \times 10^{-26} \times (0.1095 \times 10^{-9})^2$$

$$= 1.394 \times 10^{-46} \text{ kg m}^2$$

The symmetry number is 2.

$$q_r = \frac{8\pi^2 \times 1.394 \times 10^{-46} \times 1.381 \times 10^{-23} \times 300}{2 \times (6.626 \times 10^{-34})^2} = 51.9$$

$$\ln Q_r = L \ln q_r = 2.378 \times 10^{24}$$

15. The molecular vibrational partition function is

$$q_v = \frac{1}{1 - e^{-1890/T}} \cdot \frac{1}{1 - e^{-3360/T}} \left(\frac{1}{1 - e^{-954/T}}\right)^2$$

(a) $q_v = 1.002 \times 1.00 \times 1.04^2 = 1.09$

(b) $q_v = 2.14 \times 1.48 \times 3.671^2 = 42.7$

16. The Sackur-Tetrode equation is based on Eq. 14.98 for the translational partition function. This expression was obtained by replacing a summation (Eq. 14.95) by an integration (Eq. 14.96), a procedure that is valid only if the spacing between the translational levels is much smaller than $\mathbf{k}T$. This approximation is not valid at extremely low temperatures, and the Sackur-Tetrode equation then is inapplicable.

17. From the Sackur-Tetrode equation (Eq. 14.102),

$$S/\text{J K}^{-1}\text{ mol}^{-1} = 108.74 + 12.47 \ln M_r$$

For argon M_r = 39.948 and therefore

$$S/\mathrm{J\ K^{-1}\ mol^{-1}} = 108.74 + 45.98 = 154.7$$

$$S = 154.7\ \mathrm{J\ K^{-1}\ mol^{-1}}$$

18. The value of θ_v is 470 K and thus

$$q_v = \frac{1}{1 - e^{-470\ \mathrm{K}/T}}$$

(a) At T = 300 K, q_v = 1.26

(b) At T = 3000 K, q_v = 6.90

19. C_3O_2: O=C=C=C=O σ = 2

CH_4

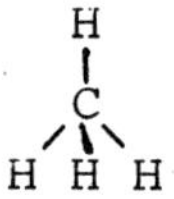

σ = 12

C_2H_6

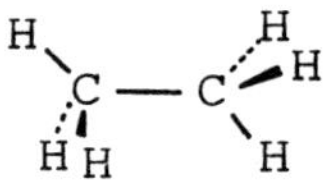

(Staggered) σ = 6

C_2H_6

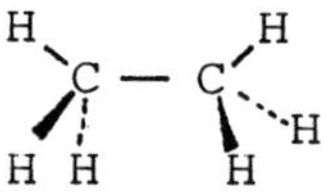

(Eclipsed) σ = 6

$CHCl_3$

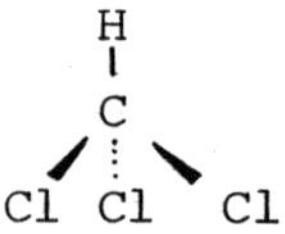

σ = 3

C_3H_6

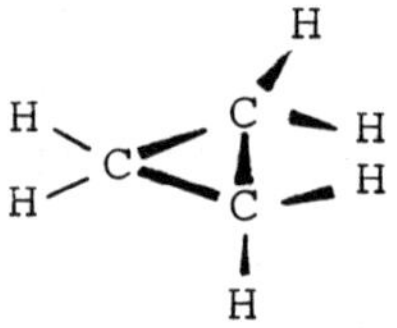

σ = 6

C_6H_6

σ = 12

NH_2D 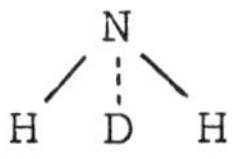 $\sigma = 1$

CH_2Cl_2 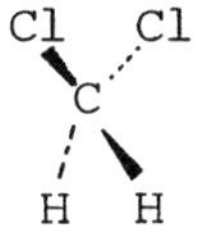 $\sigma = 2$

20. The rotational constant is

$$B = \frac{h}{B\pi^2 I}$$

$$q_r = \frac{8\pi^2 I\ \mathbf{k}T}{\sigma h^2} = \frac{\mathbf{k}T}{\sigma Bh}$$

21. We use Eq. 14.100 with $V_m = L\mathbf{k}T/P$:

$$S_m = \frac{5}{2}\ R + R\ \ln\left[\left(\frac{2\ m\ \mathbf{k}T}{h^2}\right)^{3/2} \frac{\mathbf{k}T}{P}\right]$$

For Cl_2, $m = \dfrac{2 \times 35.45 \times 10^{-3}}{6.022 \times 10^{23}} = 1.177 \times 10^{-25}$ kg

$P = 0.1$ atm $= 1.013\ 25 \times 10^4$ Pa

$$S_m = 2.5 \times 8.314 = 8.314\ \ln\left[\left(\frac{2\pi \times 1.77 \times 10^{-25}}{(6.626 \times 10^{-34})^2}\right)^{3/2} \times \frac{(1.381 \times 10^{-23} \times 298.15)^{5/2}}{1.013\ 25 \times 10^4}\right]$$

$$= 20.79 + 8.314\ \ln\ [2.186 \times 10^{63} \times 1.074 \times 10^{-55}]$$

$$= 20.79 + 160.24 = 181.0\ \text{J K}^{-1}\ \text{mol}^{-1}$$

22. The translational entropy is given by the Sackur-Tetrode equation (Eq. 14.102)

$$S_{t,m}/\text{J K}^{-1}\ \text{mol}^{-1} = 108.74 + 12.47\ \ln\ (28.01)$$

$$= 150.30$$

The rotational partition function is

$$q_r = \frac{8\pi^2 I\,\mathbf{k}T}{h^2}$$

$$= \frac{8\pi^2 \times 1.45 \times 10^{-46} \times 1.381 \times 10^{-23} \times 298.15}{(6.626 \times 10^{-34})^2}$$

$$= 107.4$$

The rotational entropy is

$$S_r = \mathbf{k}T\left(\frac{\partial \ln Q_r}{\partial T}\right)_V + \mathbf{k} \ln Q$$

$$S_{r,m} = RT\left(\frac{\partial \ln q_r}{\partial T}\right)_V + R \ln q_r \text{ since } Q_r = q_r^N$$

$$= R + R \ln q_r$$

$$S_{r,m}/\text{J K}^{-1}\text{ mol}^{-1} = 8.314 + 8.314 \ln 107.4 = 47.2$$

$$S_{r,m} = 47.2 \text{ J K}^{-1} \text{ mol}^{-1}$$

Since $\nu = 6.50 \times 10^{13}$ s^{-1}, the spacing between vibrational energy levels is

$$h\nu = 6.626 \times 10^{-34} \times 6.50 \times 10^{13} = 4.31 \times 10^{-20} \text{ J}$$

$$\frac{h\nu}{\mathbf{k}T} = 10.47 \qquad q_v = \frac{1}{1 - e^{-10.47}} \approx 1$$

The vibrational entropy is therefore negligible.

23. The fraction of molecules in the i-th level is

$$\frac{e^{-\varepsilon_i/\mathbf{k}T}}{1 + e^{-\varepsilon_1/\mathbf{k}T} + e^{-\varepsilon_2/\mathbf{k}T} + e^{-\varepsilon_3/\mathbf{k}T} + \ldots}$$

$$= \frac{e^{-\varepsilon_i/\mathbf{k}T}}{1 + e^{-\Delta\varepsilon/\mathbf{k}T} + e^{-2\Delta\varepsilon/\mathbf{k}T} + e^{-3\Delta\varepsilon/\mathbf{k}T} + \ldots}$$

$$= (1 - e^{-\Delta\varepsilon/\mathbf{k}T})\, e^{-\varepsilon_i/\mathbf{k}T}$$

The limiting value of this fraction when $T \to \infty$ is zero; this is because the molecules are now distributed evenly between an infinite number of levels.

24. (a) $S_m = \text{constant} + \frac{3}{2} R \ln M_r$ (1)

$$\frac{dS_m}{dM_r} = \frac{3}{2} \frac{R}{M_r} \quad (2)$$

(b) The heat capacity C_P is $(\partial H/\partial T)_P = (\partial S/\partial \ln T)_P$ and therefore does not depend on M_r. The Sackur-Tetrode equation can be written as

$$S_m = \text{constant} + \frac{5}{2} R \ln T \quad (3)$$

and therefore

$$C_{P,m} = \frac{5}{2} R \quad (4)$$

There is no dependence on M_r

(c) From Eq. (3)

$$\frac{dS_m}{dT} = \frac{5}{2} \cdot \frac{R}{T} \quad (5)$$

25. The standard enthalpy change for the reaction $H_2 + \frac{1}{2}O_2 \to H_2O$ at 298.15 K is the enthalpy of formation of $H_2O(g)$:

$$\Delta H^\circ_{298} = -241.82 \text{ kJ mol}^{-1}$$

$$\Delta(H^\circ_{298} - H^\circ_0) = 9.908 - 8.468 - \frac{1}{2}(8.661)$$

$$= -2.891 \text{ kJ mol}^{-1}$$

$$\Delta H^\circ_0 = -241.82 + 2.891 = -238.93 \text{ kJ mol}^{-1}$$

$$\Delta G^\circ_{298} = -238\ 930 - 298.15\ [155.52 - 102.17 - \tfrac{1}{2}(175.98)]$$

$$= -228\ 600\ \text{J mol}^{-1} = -228.6\ \text{kJ mol}^{-1}$$

26. $\Delta H^\circ_{298} = -110.53 - 241.82 + 393.51 = 41.16\ \text{kJ mol}^{-1}$

$$\Delta(H^\circ_{298} - H^\circ_0) = 8.673 + 9.908 - 8.468 - 9.364$$

$$= 0.749\ \text{kJ mol}^{-1}$$

$$\Delta H^\circ_0 = 41.16 - 0.749 = 40.41\ \text{kJ mol}^{-1}$$

$$\Delta G^\circ_{1000} = 40\ 410 - 1000\ (168.41 + 155.52 - 182.26 - 102.17) = 910\ \text{J mol}^{-1}$$

$$\ln K_P = \frac{910}{8.314 \times 1000} = -0.1095$$

$$K_P = 0.90$$

27. $\Delta H^\circ_{f,298} = -(2 \times 46.11) = -92.22\ \text{kJ mol}^{-1}$

$$\Delta(H^\circ_{298} - H^\circ_0) = (2 \times 9.916) - 8.669 - (3 \times 8.468)$$

$$= 19.832 - 8.669 - 25.404$$

$$= -14.241\ \text{kJ mol}^{-1}$$

$$\Delta H^\circ_0 = -92.220 + 14.241 = -77.98\ \text{kJ mol}^{-1}$$

(a) $\Delta G^\circ_{298} = -77\ 980 + 298.15\ [-(158.95 \times 2) + 162.42 + (3 \times 102.17)]$

$$= -77\ 980 - 94\ 782 + 48\ 426 + 91\ 386$$

$$= -32\ 950\ \text{J mol}^{-1} = -32.95\ \text{kJ mol}^{-1}$$

$$\log K_P = \frac{32\ 950}{8.314 \times 298.15} = 13.29$$

$$K_P = 5.93 \times 10^5\ \text{atm}^{-1}$$

(b) $$\Delta G^{\circ}_{1000} = -77\ 980 + [-(2 \times 203.47) + 197.95 + (3 \times 136.98)]1000$$

$$= -77\ 980 - 406\ 940 + 197\ 950 + 410\ 940$$

$$= 123\ 970 \text{ J mol}^{-1} = 123.97 \text{ kJ mol}^{-1}$$

$$\log K_P^u = -\frac{123\ 970}{8.314 \times 1000} = -14.91$$

$$K_P = 3.34 \times 10^{-7} \text{ atm}^{-2}$$

28. The symmetry numbers are given below the molecules (they are 1 for atoms) and the statistical factors are shown above and below the arrows; for simplicity ^{35}Cl is written as Cl and ^{37}Cl as Cl*

(a) $\underset{2}{Cl-Cl} + Cl^* \underset{1}{\overset{2}{\rightleftharpoons}} \underset{1}{Cl-Cl^*} + Cl$ $\quad K = 2$

(b) $\underset{2}{Cl-Cl} + \underset{2}{Cl^*-Cl^*} \underset{1/2}{\overset{2}{\rightleftharpoons}} 2\underset{1}{Cl-Cl^*}$ $\quad K = 4$

(c) $\underset{12}{CCl_4} + Cl^* \underset{1}{\overset{4}{\rightleftharpoons}} \underset{3}{CCl^*Cl_3} + Cl$ $\quad K = 4$

(d) $\underset{3}{NCl_3} + Cl^* \underset{1}{\overset{3}{\rightleftharpoons}} \underset{1}{NCl^*Cl_2} + Cl$ $\quad K = 3$

(e) $\underset{2}{Cl_2O} + Cl^* \underset{1}{\overset{2}{\rightleftharpoons}} \underset{1}{ClCl^*O} + Cl$ $\quad K = 2$

In each case K is $\sigma_A\sigma_B/\sigma_Y\sigma_Z$ and is l/r.

29. The fraction is

$$\frac{dN}{N} = \frac{B\ e^{-mu_x^2/2\mathbf{k}T}\ du_x}{B\int_0^{\infty} e^{-mu_x^2/2\mathbf{k}T}\ du_x}$$

From the appendix to Chapter 14 the integral is

$$\frac{1}{2}\left(\frac{2\pi\mathbf{k}T}{m}\right)^{1/2}$$

Thus

$$\frac{dN}{N} = \left(\frac{2m}{\pi \mathbf{k}T}\right)^{1/2} e^{-mu_x^2/2\mathbf{k}T} du_x$$

A plot of $(dN/N)/du_x$ against u_x is symmetrical about the $(dN/N)/du_x$ axis; the maximum is therefore at $u_x = 0$, which is the most probable speed.

30. $\varepsilon_x = \frac{1}{2} mu_x^2$

and therefore $\frac{d\varepsilon_x}{du_x} = (2\varepsilon_x m)^{1/2}$

From the expression obtained in Problem 29,

$$\frac{dN_x}{N} = \left(\frac{2m}{\pi \mathbf{k}T}\right)^{1/2} e^{-\varepsilon_x/\mathbf{k}T} \frac{1}{(2\varepsilon_x m)^{1/2}} d\varepsilon_x$$

$$= (\pi \varepsilon_x \mathbf{k}T)^{-1/2} e^{-\varepsilon_x/\mathbf{k}T} d\varepsilon_x$$

The average one-dimensional energy is

$$\bar{\varepsilon}_x = \int_0^\infty \varepsilon_x \frac{dN}{N} = (\pi \mathbf{k}T)^{-1/2} \int_0^\infty \varepsilon_x^{1/2} e^{-\varepsilon_x/\mathbf{k}T} d\varepsilon_x$$

$$= (\pi \mathbf{k}T)^{1/2} \frac{1}{2} \mathbf{k}T (\pi \mathbf{k}T)^{-1/2}$$

$$= \frac{1}{2} \mathbf{k}T$$

31. From Eq. 14.36 and 14.37, with $\beta = 1/\mathbf{k}T$ and $u^2 = u_x^2 + u_y^2$,

$$dP_x dP_y = B^2 e^{-mu^2/2\mathbf{k}T} du_x du_y$$

We consider a circular shell of radius u and replace $du_x du_y$ by $2\pi u\ du$. Then

$$dP = 2\pi B^2 e^{-mu^2/2\mathrm{k}T} u\ du$$

and therefore

$$\frac{dN}{N} = \frac{e^{-mu^2/2\mathbf{k}T}\ u\ du}{\int_0^\infty e^{-mu^2/2\mathbf{k}T}\ u\ du}$$

$$= \frac{e^{-mu^2/2\mathbf{k}T}\ u\ du}{\frac{1}{2} \cdot \frac{2\mathbf{k}T}{m}}$$

$$= \frac{m}{\mathbf{k}T}\ e^{-mu^2/2\mathbf{k}T}\ u\ du$$

$$\varepsilon = \frac{1}{2}\ mu^2 \text{ and } \frac{d\varepsilon}{du} = (2\varepsilon m)^{1/2}$$

Thus

$$\frac{dN}{N} = \frac{m}{\mathbf{k}T}\ e^{-\varepsilon/\mathbf{k}T}\left(\frac{2\varepsilon}{m}\right)^{1/2} \frac{d\varepsilon}{(2\varepsilon m)^{1/2}}$$

$$= \frac{1}{\mathbf{k}T}\ e^{-\varepsilon/\mathbf{k}T}\ d\varepsilon$$

The fraction with energy in excess of ε^* is

$$\frac{N^*}{N} = \frac{1}{\mathbf{k}T}\int_{\varepsilon^*}^\infty e^{-\varepsilon/\mathbf{k}T}\ d\varepsilon = -\ e^{-\varepsilon/\mathbf{k}T}\Big|_{\varepsilon^*}^{\infty} = e^{-\varepsilon^*/\mathbf{k}T}$$

32. The partition function is

$$Q = 1 + e^{-\varepsilon/\mathbf{k}T}$$

(a) The molar internal energy is (Table 14.3, p. 666)

$$U_m - U_{m,0} = RT^2\left(\frac{\partial \ln Q}{\partial T}\right)_V$$

$$\left(\frac{\partial \ln Q}{\partial T}\right)_V = \frac{1}{Q}\left(\frac{\partial Q}{\partial T}\right)_V = \frac{\varepsilon}{Q\mathbf{k}T^2}\ e^{-\varepsilon/\mathbf{k}T}$$

$$= \frac{L\varepsilon}{QRT^2}\ e^{-\varepsilon/\mathbf{k}T}$$

Then

$$U_m - U_{m,0} = L\varepsilon \frac{e^{-\varepsilon/\mathbf{k}T}}{1 + e^{-\varepsilon/\mathbf{k}T}}$$

$$= \frac{1}{2} L\varepsilon \quad \text{when } T \to \infty$$

(Under these conditions half of the molecules are in the ground state and half in the state of energy ε).

(b) The molar entropy is (see Eq. 14.80)

$$S_m = \frac{U_m - U_{m,0}}{T} + R \ln Q$$

When $T \to \infty$ the first term approaches zero and $Q \to 2$; thus

$$S_m(T \to \infty) = R \ln 2$$

(c) $H - U_0 = U - U_0 + N\mathbf{k}T = \frac{1}{2} N\varepsilon + N\mathbf{k}T$

$H_m - U_{0,m} = RT$ when T is very large

(d) $G - U_0 = -N\mathbf{k}T \ln q + N\mathbf{k}T$

$G_m - U_{0,m} = -R \ln 2 + RT = RT$ when T is very large.

33. For two-dimensional translation motion

$$q_t = \frac{2\pi m\, \mathbf{k}T\, A}{h^2}$$

$$\ln Q = N \ln \left(\frac{e}{N}\right) + N = N \ln\left[\frac{A}{Nh^2} 2\pi m\, \mathbf{k}T\right] + N$$

$$S = \mathbf{k}T\left(\frac{\partial \ln Q}{\partial T}\right)_A + \mathbf{k} \ln Q$$

$$S_m = RT \cdot \frac{1}{T} + R + R \ln\left[\frac{A}{Nh^2} \cdot 2\pi m\, \mathbf{k}T\right]$$

$$= 2\, R + R \ln\left(\frac{2\pi m\, \mathbf{k}T\, A}{Nh^2}\right)$$

For Ar, $m = \dfrac{39.948 \times 10^{-3}}{6.022 \times 10^{23}} = 6.624 \times 10^{-26}$ kg

If 10^{10} molecules are adsorbed on an area of 1 cm^2 at 25°C,

$$S_m = 2 \times 8.314 + 8.314 \ln$$

$$\left[\frac{2 \times 6.634 \times 10^{-26} \times 298.15 \times 1.381 \times 10^{23} \times 10^{-4}}{10^{10} \times (6.626 \times 10^{-34})^2}\right]$$

$$= 16.63 + 8.314 \ln (3.909 \times 10^7)$$

$$= 162.0 \text{ J K}^{-1} \text{ mol}^{-1}$$

34. Mass of I atom, $m = 126.90 \times 10^{-3}/6.022 \times 10^{23}$

$$= 2.107 \times 10^{-25} \text{ kg}$$

Translational partition function for the I atom, with $V = 1$ m^3, is

$$q_t(\text{I}) = \frac{(2\pi \times 2.107 \times 10^{-25} \times 1.381 \times 10^{-23})^{3/2}}{(6.626 \times 10^{-34})^3} \times$$

$$(1273.15)^{3/2} = 1.221 \times 10^{34}$$

The degeneracy of the ground state is

$$2(\tfrac{3}{2}) + 1 = 4; \text{ thus}$$

$$q_t(\text{I}) = 4 \times 1.221 \times 10^{34} = 4.884 \times 10^{34}$$

For the iodine molecule,

$$q_t(\text{I}_2) = \frac{(2\pi \times 2 \times 2.107 \times 10^{-25} \times 1.381 \times 10^{-23})^{3/2}}{(6.626 \times 10^{-34})^3} \times$$

$$(1273.15)^{3/2}$$

$$= 2^{3/2} \, q_t(\text{I}) = 3.453 \times 10^{34} \text{ m}^{-3}$$

$$q_r(\text{I}_2) = \left[\frac{8\pi^2 \times 7.426 \times 10^{-45} \times 1.381 \times 10^{-23} \times 1273.15}{2(6.626 \times 10^{-34})^2}\right]$$

$$= 1.174 \times 10^4$$

$$q_v(I_2) = \frac{1}{1 - \exp\left(\frac{-6.626 \times 10^{-34} \times 21\ 367.0 \times 2.998}{10^{-8} \times 1.381 \times 10^{-23} \times 1273.15}\right)}$$

$$= \frac{1}{1 - 0.786} = 4.66$$

The molecular partition function for I_2 is thus

$$q_{I_2} = 3.453 \times 10^{34} \times 1.174 \times 10^4 \times 4.66$$

$$= 1.890 \times 10^{39}$$

The molecular equilibrium constant K is thus

$$\frac{(4.884 \times 10^{34})^2}{1.890 \times 10^{39}} e^{-148\ 450/8.314 \times 1273.15}$$

$$= 1.263 \times 10^{30} \times 8.113 \times 10^{-7} = 1.025 \times 10^{24}\ m^{-3}$$

Its value in molar units is

$$K_c = 1.701\ mol\ m^{-3} = 1.701 \times 10^{-3}\ mol\ dm^{-3}$$

At 1273.15 K, 1 mol dm^{-3} = 104.46 atm

$$K_p = 0.178\ atm$$

(The experimental value obtained by Starck and Bodenstein in 1910 was 0.165 atm).

35. Mass of Na atom = $22.99 \times 10^{-3}/6.022 \times 10^{23}$

$$= 3.818 \times 10^{-26}\ kg$$

The electronic partition function = $2(\frac{1}{2}) + 1 = 2$

The molecular partition function for Na is

$$q(Na) = \frac{2 \times (2\pi \times 3.818 \times 10^{-26} \times 1.381 \times 10^{-23})^{3/2}}{(6.626 \times 10^{-34})^3} \times (1000)^{3/2}$$

$$= 1.311 \times 10^{33}\ m^{-3}$$

For Na_2,

$$q_t(Na_2)/m^{-3} = \frac{(2\pi \times 2 \times 3.818 \times 10^{-26})^{3/2}}{(6.626 \times 10^{-34})^3} \times (1.381 \times 10^{-23} \times 1000)^{3/2}$$

$$= 1.854 \times 10^{33} \quad (= q_t(Na) \times \sqrt{2})$$

$$\text{Moment of inertia of } Na_2 = \frac{3.818 \times 10^{-26}}{2} \times (0.3716 \times 10^{-9})^2$$

$$= 2.636 \times 10^{-45}\ kg\ m^2$$

$$q_r(Na_2) = \frac{8\pi^2 \times 2.636 \times 10^{-45} \times 1.381 \times 10^{-23} \times 1000}{2 \times (6.626 \times 10^{-34})^2}$$

$$= 3274$$

$q_v(Na_2)$

$$= \frac{1}{1 - \exp\left(\dfrac{-159.2 \times 2.998 \times 10^{10} \times 6.626 \times 10^{-34}}{1.381 \times 10^{-23} \times 1000}\right)}$$

$$= 4.89$$

Thus the partition function for Na_2 at 1000 K

$$= 1.854 \times 10^{33} \times 3274 \times 4.89$$

$$= 2.97 \times 10^{37}\ m^{-3}$$

The molecular equilibrium constant is

$$K = \frac{(1.311 \times 10^{33})^2}{2.97 \times 10^{37}}\ e^{-704\ 000/8.314 \times 1000}$$

$$= 5.787 \times 10^{28} \times 2.101 \times 10^{-4}$$

$$= 1.216 \times 10^{25}\ m^{-3}$$

$K_c = 20.2 \text{ mol m}^{-3} = 0.0202 \text{ mol dm}^{-3}$

At 1000 K, 1 mol $\text{dm}^{-3} = 82.05$ atm

$K_p = 1.66$ atm

36. Mass of Cl atom = $35.45 \times 10^{-3}/6.022 \times 10^{23}$ kg

$= 5.89 \times 10^{-26}$ kg

Translational partition function for the Cl atom, with $V = 1 \text{ m}^3$, is

$$Q_t(\text{Cl}) = \frac{(2\pi \times 5.89 \times 10^{-26} \times 1.381 \times 10^{-23} \times 1200)^{3/2}}{(6.626 \times 10^{-34})^3}$$

$= 1.651 \times 10^{33}$

The degeneracy of the $^2P_{3/2}$ state is 4; that of the $^2P_{1/2}$ state is 2; the electronic partition function is thus

$Q_e(\text{Cl}) = 4 + 2e^{-\varepsilon/\mathbf{k}T}$

$$= 4 + 2\exp\left(\frac{-881 \times 2.998 \times 10^{10} \times 6.626 \times 10^{-34}}{1.381 \times 10^{-23} \times 1200}\right)$$

$= 4 + 2e^{-1.056} = 4.696$

The complete partition function for the Cl atom is thus

$Q(\text{Cl}) = 7.753 \times 10^{33}$

For the Cl_2 molecule,

$q_t(\text{Cl}_2) = 2^{3/2} q_t(\text{Cl})/\text{m}^3 = 4.70 \times 10^{33}$

The moment of inertia of Cl_2 is

$$I = \mu r^2 = \frac{1}{2} m_{\text{Cl}} r^2 = \frac{1}{2} \times 5.89 \times 10^{-26} \times (1.99 \times 10^{-10})^2 \text{ kg m}^2$$

$$= 1.167 \times 10^{-45} \text{ kg m}^2$$

The rotational partition function of Cl_2 ($\sigma = 2$) is

$$q_r(Cl_2) = \frac{8\pi^2 \times 1.167 \times 10^{-45} \times 1.381 \times 10^{-23} \times 1200}{2 \times (6.626 \times 10^{-34})^2}$$

$$= 1739$$

The vibrational partition function is

$$q_v(Cl_2) = \frac{1}{1 - \exp\left(\frac{-565 \times 2.998 \times 10^{10} \times 6.626 \times 10^{-34}}{1.381 \times 10^{-23} \times 1200}\right)}$$

$$= 2.033$$

The molecular partition function for Cl_2 is thus

$$q(I_2) = 4.70 \times 10^{33} \times 1739 \times 2.033 = 1.66 \times 10^{37}$$

The molecular equilibrium constant is thus

$$\frac{(7.753 \times 10^{33})^2}{1.66 \times 10^{37}} e^{-240\ 000/8.314 \times 1200}$$

$$= 3.62 \times 10^{30} \times 3.57 \times 10^{-11} = 1.29 \times 10^{20} \text{ m}^{-3}$$

Its value in molar units is

$$K_c = 2.14 \times 10^{-4} \text{ mol m}^{-3} = 2.14 \times 10^{-7} \text{ mol dm}^{-3}$$

At 1200 K, 1 mol dm^{-3} = 98.5 atm

$$K_p = 2.11 \times 10^{-5} \text{ atm}$$

37. $\partial\varepsilon^2 = \langle\varepsilon^2\rangle - \langle\varepsilon\rangle^2$

$$= \sum_i P_i \varepsilon_i^2 - (\sum_i P_i \varepsilon_i)^2$$

$$= \frac{\sum_i \varepsilon_i^2 e^{-\beta\varepsilon_i}}{q} - \frac{(\sum_i \varepsilon_i e^{-\beta\varepsilon_i})^2}{q^2}$$

$$= \frac{\sum_i \left(\frac{\partial^2}{\partial \beta^2} e^{-\beta \varepsilon_i}\right)_V}{q} - \frac{\sum_i \left(\frac{\partial}{\partial \beta} e^{-\beta \varepsilon_i}\right)_V^2}{q^2}$$

$$= \frac{1}{q}\left(\frac{\partial^2 q}{\partial \beta^2}\right)_V - \frac{1}{q^2}\left(\frac{\partial q}{\partial \beta}\right)_V^2 \qquad (1)$$

The partition function for a harmonic oscillator is

$$q_v = \frac{1}{1 - e^{-h\nu/\mathbf{k}T}} = \frac{1}{1 - e^{-\beta h\nu}} \qquad (2)$$

Then

$$\frac{dq_v}{d\beta} = -\frac{h\nu\ e^{-\beta h\nu}}{(1 - e^{-\beta h\nu})^2} \qquad (3)$$

and

$$\frac{d^2 q_v}{d\beta^2} = \frac{(h\nu)^2\ e^{-\beta h\nu}(1 + e^{-\beta h\nu})}{(1 - e^{-\beta h\nu})^3} \qquad (4)$$

Substitution of these two expressions into Eq. (1) gives, after some reduction,

$$\partial \varepsilon_v^2 = \frac{(h\nu)^2\ e^{-\beta h\nu}}{(1 - e^{-\beta h\nu})^2}$$

$$\partial \varepsilon_v = \frac{h\nu\ e^{-\beta h\nu/2}}{1 - e^{-\beta h\nu}}$$

$$= \frac{h\nu}{e^{h\nu/2\mathbf{k}T} - e^{-h\nu/2\mathbf{k}T}}$$

As $T \to \infty$ the exponentials can be expanded and only the first term accepted:

$$\partial \varepsilon_v = \frac{h\nu}{(1 + \frac{h\nu}{2\mathbf{k}T}) - (1 - \frac{h\nu}{2\mathbf{k}T})}$$

$$= \mathbf{k}T$$

CHAPTER 15: WORKED SOLUTIONS
THE SOLID STATE

1. (a) A unit cell has 8 lattice points at the corners of a cube; each of them are shared with 7 other unit cells. Therefore, only $\frac{1}{8}$ of the 8 belong to a particular fcc cell. Each face has an additional lattice point shared between two cells; there are therefore $\frac{1}{2} \times 6 = 3$ lattice points in the faces. For the unit cell: 1 (from corners) + 3 (from faces) = 4 lattice points.

   (b) A bcc lattice has 1 lattice point belonging entirely to the unit cell, plus $\frac{1}{8} \times 8$ corner points. There are thus 1 + 1 = 2 lattice points.

2. (a) The end-centered lattice has $\frac{1}{8} \times 8 + \frac{1}{2} \times 2 = 2$ lattice points. Since one basis is at each lattice point, each unit cell has 2 basis groups.

   (b) The primitive lattice has one lattice point and there is therefore only one basis group.

3. (a) Consider the array of circles

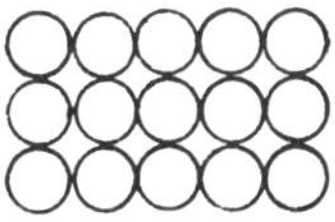

The area belonging to each circle is shown as a dotted box of area $4R^2$. The area of the circle is πR^2. The efficiency of filling space is $\pi R^2/4R^2 = 78.5\%$.

(b) Circles on a triangular lattice are shown below:

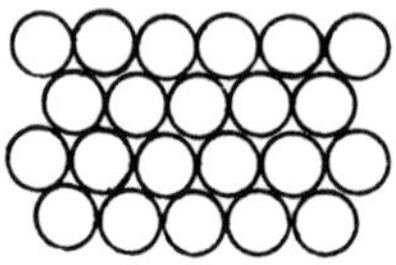

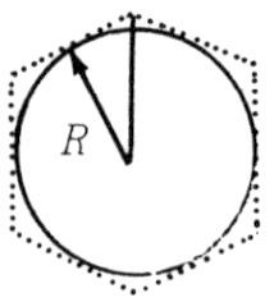

The hexagonal area belonging to a single circle is shown by the dotted lines. The hexagonal area is made up of twelve right triangles, each having an area of

$$(\tfrac{1}{2}R)(\tfrac{1}{\sqrt{3}}R) = (\tfrac{1}{2\sqrt{3}})R^2$$

The total area is $12(1/2\sqrt{3})R^2 = 2\sqrt{3}R^2$ and the efficiency of filling space is given by

$$\pi R^2/2\sqrt{3}\ R^2 = \frac{\pi}{2\sqrt{3}} = 90.7\%$$

(c) The triangular form is more efficient by a factor of $\frac{90.7}{78.5} = 1.16$.

4. The volume of a unit cell with right angles is the product abc of its edges. Since one mole of the crystal contains L/z unit cells, the molar volume is $V = \frac{abc\ L}{z}$. The molar mass M divided by V is the density ρ, and substitution gives $\rho = \frac{Mz}{abcL}$

5. From problem 4: $\rho = \frac{Mz}{abcL}$

Molar mass of $[BrC_6H_4C(F){=}]_2$ = 373.94 g mol^{-1}

$$\rho = \frac{4(373.94\ \text{g mol}^{-1})}{(28.32)(7.36)(6.08)\ \times 10^{-24}\ \text{cm}^3 \times 6.022 \times 10^{23}}$$

$= 1.96\ g\ cm^{-3}$

6. From Problem 4: $\rho = \frac{Mz}{abcL}$

For a face-centered cubic lattice all edge lengths are equal, that is $a = b = c$.

$$z = \frac{\rho\ abcL}{M}$$

$$= \frac{2.328\ 99\ g\ cm^{-1}(5.431\ 066)^3(10^{-8}\ cm^3)(6.022 \times 10^{23})}{28.085\ 41\ g}$$

≈ 8

7. $$\rho = \frac{Mz}{abcL} = \frac{4(58.45\ g\ mol^{-1})}{(5.629)^3(10^{-8})^3(6.022 \times 10^{23}\ mol^{-1})}$$

$$= \frac{233.8\ g}{178.4 \times 0.6022\ cm^3} = 2.176\ g\ cm^{-3}$$

CRC value = $2.165\ g\ cm^{-3}$ at 25°C. The lower density given in the handbook may be due to voids and other imperfections in the crystal.

8. The molar mass of KCl is $74.55\ g\ mol^{-1}$

$$\rho = \frac{Mz}{abcL} = \frac{4(74.55\ g\ mol^{-1})}{(6.278 \times 10^{-8}\ cm)^3(6.022 \times 10^{23}\ mol^{-1})}$$

$= 2.001\ g\ cm^{-3}$

CRC value: $1.984\ g\ cm^{-3}$

9. There are 4 Ca^{2+} ions per unit cell and 8 associated F^- ions. Consequently, $z = 4$ and rearrangement of

$$\rho = \frac{Mz}{abcL} \text{ with } a = b = c \text{ gives}$$

$$a^3 = zM/\rho L = \frac{4(78.08 \text{ g mol}^{-1})}{(3.18 \text{ g cm}^{-3})(6.022 \times 10^{23} \text{ mol}^{-1})}$$

$$= 1.63 \times 10^{-22} \text{ cm}^3$$

$$a = 5.46 \times 10^{-8} \text{ cm} = 5.46 \text{ Å} = 546 \text{ pm}$$

10. The intercepts along the axes are spaced at a/h, b/k, c/l. For a cubic system $1a = \frac{a}{h}$, , $h = 1$; $\frac{a}{2} = \frac{a}{k}$, $k = 2$; $\frac{2a}{3} = \frac{a}{l}$, $l = \frac{3}{2}$. Clearing fractions we have (hkl) $= (243)$.

11. The faces are shown below:

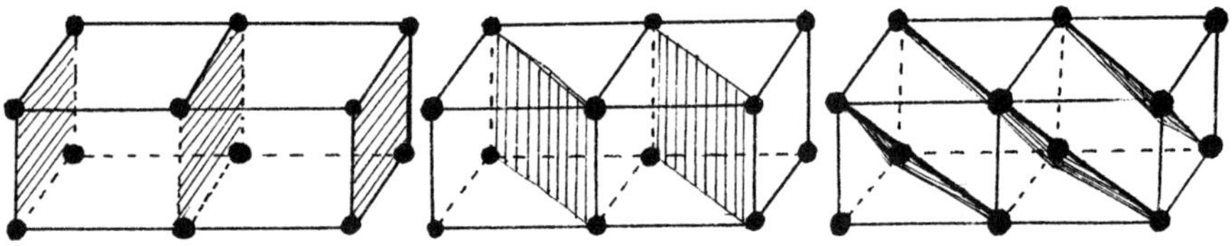

100 planes 110 planes 111 planes

The spacings are calculated from the formula $d_{hkl} = a/(h^2 + k^2 + l^2)^{1/2}$ or from trigonometry as demonstrated from the 110 planes:

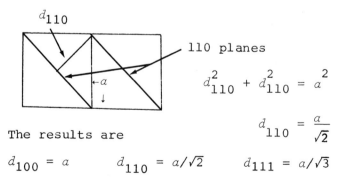

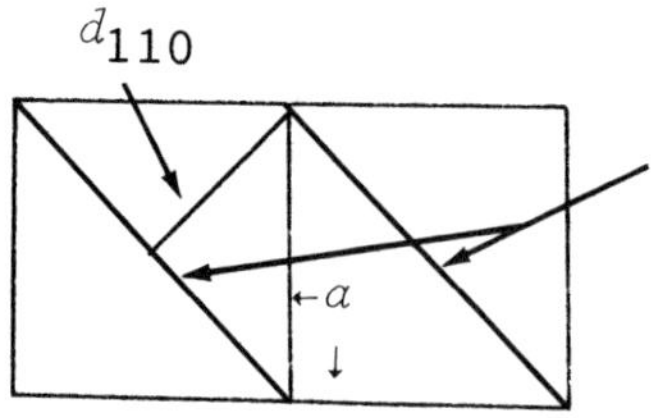

110 planes

$$d_{110}^2 + d_{110}^2 = a^2$$

$$d_{110} = \frac{a}{\sqrt{2}}$$

The results are

$d_{100} = a$ $d_{110} = a/\sqrt{2}$ $d_{111} = a/\sqrt{3}$

12.

	Originals	Reciprocals	Miller indices
(a)	$(2a, b, 3c)$	$\frac{1}{2}, 1, \frac{1}{3}$	(362)
(b)	$(2a, -3b, 2c)$	$\frac{1}{2}, -\frac{1}{3}, \frac{1}{3}$	$(3\bar{2}3)$
(c)	$(a, b, -c)$	1, 1, -1	$(11\bar{1})$

13. $$d_{hkl} = \frac{a}{(h^2 + k^2 + l^2)^{1/2}}$$

$$d_{100} = \frac{389 \text{ pm}}{1} = 389 \text{ pm}$$

$$d_{111} = \frac{389 \text{ pm}}{(1^2 + 1^2 + 1^2)^{1/2}} = \frac{389 \text{ pm}}{\sqrt{3}} = 225 \text{ pm}$$

$$d_{121} = \frac{389 \text{ pm}}{(1^2 + 2^2 + (-1)^2)^{1/2}} = \frac{389 \text{ pm}}{\sqrt{6}} = 159 \text{ pm}$$

14. $\lambda = 2d_{hkl} \sin \theta \quad \lambda = 154$ pm

From Eq. 15.5

$$\frac{1}{d^2} = \frac{h^2}{a^2} + \frac{k^2}{b^2} + \frac{l^2}{c^2}$$

$$1/d_{100}^2 = \frac{1}{a^2}; \; d_{100} = a = 488 \text{ pm}$$

$$1/d_{010}^2 = \frac{1}{b^2}; \; d_{010} = b = 666 \text{ pm}$$

$$1/d_{111}^2 = \frac{1}{a^2} + \frac{1}{b^2} + \frac{1}{c^2} = \frac{1}{488^2} + \frac{1}{666^2} + \frac{1}{832^2}$$

$$= 4.199 \times 10^{-6} + 2.254 \times 10^{-6} + 1.445 \times 10^{-6}$$

$$= 7.898 \times 10^{-6}$$

$$d_{111} = 356 \text{ pm}$$

$$\sin \theta = \lambda/2d_{100} = 154 \text{ pm}/2(488 \text{ pm})$$

$$= 0.1578$$

$$\theta_{100} = 9.08°$$

$$\sin\theta = \lambda/2d_{010} = 154 \text{ pm}/2(666 \text{ pm})$$

$$= 0.1156$$

$$\theta_{010} = 6.64°$$

$$\sin\theta = \lambda/2d_{111} = 154 \text{ pm}/2(356 \text{ pm})$$

$$= 0.2163$$

$$\theta_{111} = 12.49°$$

15. $\lambda = 2d_{hkl} \sin\theta$ $\qquad\lambda = 45.5$ pm

From Eq. 15.5

$$\frac{1}{d_{hkl}^2} = \frac{h^2}{a^2} + \frac{k^2}{b^2} + \frac{l^2}{c^2}$$

$$1/d_{100}^2 = 1/a^2;\ d_{100} = 482 \text{ pm}$$

$$1/d_{010}^2 = \frac{1}{b^2};\ d_{010} = 684 \text{ pm}$$

$$1/d_{111}^2 = \frac{1}{a^2} + \frac{1}{b^2} + \frac{1}{c^2} = \frac{1}{482^2} + \frac{1}{684^2} + \frac{1}{867^2}$$

$$= 4.304 \times 10^{-6} + 2.137 \times 10^{-6} + 1.330 \times 10^{-6}$$

$$= 7.771 \times 10^{-6}$$

$$d_{111} = 359 \text{ pm}$$

$$\sin\theta = \lambda/2d_{100} = 45.5 \text{ pm}/2(482 \text{ pm})$$

$$= 0.0472$$

$$\theta_{100} = 2.70°$$

$$\sin\theta = \lambda/2d_{010} = 45.5 \text{ pm}/2(684 \text{ pm})$$

$$= 0.0333$$

$$\theta_{010} = 1.91°$$

$$\sin\theta = \lambda/2d_{111} = 45.5\ \text{pm}/2(359\ \text{pm})$$

$$= 0.0634$$

$$\theta_{111} = 3.63°$$

16. $\lambda = 2d_{hkl}\sin\theta$ $\lambda = 154$ pm

From Eq. 15.4

$$\frac{1}{d_{hkl}^2} = \frac{h^2 + k^2}{a^2} + \frac{l^2}{c^2}$$

$$1/d^2_{(100)} = \frac{1}{a^2} = \frac{1}{(967\ \text{pm})^2}$$

$$d_{100} = 967\ \text{pm}$$

$$1/d^2_{111} = \frac{1^2 + 1^2}{(967)^2} + \frac{1^2}{(892)^2}$$

$$= 2.139 \times 10^{-6} + 1.257 \times 10^{-6} = 3.396 \times 10^{-6}$$

$$d_{111} = 543\ \text{pm}$$

$$\sin\theta = \lambda/2d_{100} = 154\ \text{pm}/2(967\ \text{pm})$$

$$= 0.0796$$

$$\theta_{100} = 4.57°$$

$$\sin\theta = \lambda/2d_{111} = 154/2(543\ \text{pm})$$

$$= 0.1418$$

$$\theta_{111} = 8.15°$$

17. First find the distance along the two axis to the point of intersection:

	A	*B*	*C*	*D*
xa, yb, ∞c:	∞, 1, ∞	-1, 1, ∞	1, $\frac{1}{2}$, ∞	3, 2, ∞

The reciprocals after clearning fractions are;

(0, 1, 0)   (-1, 1, 0)   (2, 1, 0)   (1, 1, 0)

18. (a) Let $n = 1$ in the Bragg equation $n\lambda = 2d\sin\theta$

Then $\theta = \sin^{-1}(\lambda/2d)$

$$= \sin^{-1}(70.8/650) = \sin^{-1}(0.1089) = 6.25°$$

(b) $\theta = \sin^{-1}(\lambda/2d) = \sin^{-1}(154/650)$

$$= \sin^{-1}(0.2369) = 13.70°$$

Notice that the shorter the wavelength, the smaller the diffraction angle.

19. The kinetic energy of the electron is $\frac{1}{2}mu^2$ and is also VQ; it thus follows that

$$u = \left(\frac{2VQ}{m}\right)^{1/2}$$

$$= \left[\frac{2(40 \times 10^3\ \text{V})(1.60 \times 10^{-19}\ \text{C})}{9.11 \times 10^{-31}\ \text{kg}}\right]^{1/2}$$

$$= (14.5 \times 10^{15}\ \text{J kg}^{-1})^{1/2}$$

Since $\text{J} = \text{kg}\cdot\text{m}^2\cdot\text{s}^{-2}$

$$u = 1.19 \times 10^8\ \text{m s}^{-1}$$

Substitution into the de Broglie equation (11.56) gives

$$\lambda = \frac{6.63 \times 10^{-34}\ \text{J s}}{9.11 \times 10^{-31}\ \text{kg}(1.19 \times 10^8\ \text{m s}^{-1})}$$

$$= 0.061 \times 10^{-10}\ \text{m} = 6.1\ \text{pm}$$

20. From Figure 15.19 it is determined that the face-centered cubic (*fcc*) system is the only one that conforms to the data. Note that the symmetry of

the crystal determines which indices will appear.

21. From Figure 15.19, it can only be *bbc*.

22. (a) From Figure 15.19 it is seen that the ratio 7 is not allowed for cubic systems. The ratio must be 2, 4, 6, etc. Consequently the structure is *bcc*.

(b) For a *bcc* system, $a = b = c$, $z = 2$, and from Problem 4, $\rho = zA/a^3L$ or

$$a^3 = 2(55.85 \text{ g mol}^{-1})/(7.90 \text{ g cm}^{-3}) \times (6.022 \times 10^{23} \text{ mol}^{-1}) = 2.348 \times 10^{-23}$$

$$a = 2.86 \times 10^{-8} \text{ cm} = 286 \text{ pm}$$

For 100 type planes, d is $a/2$ since the planes are actually (200). Therefore from $2d\sin\theta = n\lambda$, $\sin\theta = \dfrac{154.18 \text{ pm}}{286 \text{ pm}} = 0.539$

$$\theta = 32.6°$$

(c) The body diagonal is the smallest interatomic distance and has the value $\dfrac{\sqrt{3}a}{2}$. Therefore the radius is the distance from the center of one Fe atom to the center of the central atom divided by 2:

$$r_{Fe} = \frac{\sqrt{3}a}{2.2} = \frac{1.732(286)}{4} = 123.8 \text{ pm}$$

23. There are two atoms in a body-centered lattice and we may write:

$$\text{density} = \frac{(\text{number of atoms/cell})(\text{atomic mass})}{La^3}$$

$$0.856 \text{ g cm}^{-3} = \frac{2(39.102 \text{ g mol}^{-1})}{(6.022 \times 10^{23} \text{ mol}^{-1})a^3}$$

$$a^3 = \frac{2 \times 39.102}{6.022 \times 10^{23} \times 0.856} = 1.517 \times 10^{-22} \text{ cm}^{-3}$$

$$a = 5.333 \times 10^{-8} \text{ cm} = 0.5333 \text{ nm}$$

$$= 533.3 \text{ pm}$$

Then from the equation

$$d_{hkl} = \frac{a}{h^2 + k^2 + l^2} = \frac{533.3 \text{ nm}}{h^2 + k^2 + l^2}$$

For (200) planes, d_{200} = 533.3/ 4 = 266.7 pm

For (110) planes, d_{110} = 533.3/ 2 = 377.1 pm

For (222) planes, d_{222} = 533.3/ 12 = 154.0 pm

24. First determine the d values for the three lines and take their ratios:

$$2d_1 = \frac{\lambda}{\sin 14.18}$$

$$2d_2 = \frac{\lambda}{\sin 20.25}$$

$$2d_3 = \frac{\lambda}{\sin 25.10}$$

$$d_1 : d_2 : d_3 = \frac{1}{\sin 14.18} : \frac{1}{\sin 20.25} : \frac{1}{\sin 25.10}$$

$$= 4.082 : 2.889 : 2.357$$

$$= 1 : 0.7077 : 0.577$$

From Problem 10 for the cubic lattice

$$d_{100} = a$$

$$d_{110} = \frac{\sqrt{2}}{2}\, a = 0.707$$

$$d_{111} = \frac{\sqrt{3}}{3}\, a = 0.5773$$

The ratios thus correspond to the cubic structure. To confirm the structure, if K^+ and Cl^- reflect equally $d = a/2$, and the theoretical density could be compared to the experimental value.

25. $$d_{111} = \frac{\lambda}{2 \sin \theta} = \frac{154.18 \text{ pm}}{2 \sin 19.076} = 235.9 \text{ pm}$$

$$= \frac{a}{(h^2 + k^2 + l^2)^{1/2}} = 2.359 \times 10^{-10} \text{ m}$$

$$\frac{a}{(1 + 1 + 1)^{1/2}} = \frac{a}{(3)^{3/2}} = 2.359 \times 10^{-10} \text{ m}$$

Then the effective volume of each Ag atom is

$$V_{Ag} = M/\rho = \frac{0.107\ 87 \text{ kg mol}^{-1}}{10\ 500 \text{ kg m}^{-3}\ 6.022 \times 10^{23} \text{ mol}^{-1}}$$

$$= 1.706 \times 10^{7} \text{ pm}^3$$

$$V_{cell} = a^3 = (408.6 \text{ pm})^3 = 6.82 \times 10^{7} \text{ pm}^3$$

The number of atoms per unit cell is

$$N = \frac{V \text{ cell}}{V_{Ag}} = \frac{6.82 \times 10^{7} \text{ pm}^3}{1.706 \times 10^{7} \text{ pm}^3} = 4.0$$

This is an indication that Ag is *fcc*.

26. $$d_{111} = \frac{\lambda}{2 \sin \theta} = \frac{154.2 \text{ pm}}{2 \sin 16.72} = 268.0 \text{ pm}$$

$$d_{111} = 268.0 \text{ pm} = \frac{a}{(h^2 + k^2 + l^2)^{1/2}} = \frac{a}{(3)^{1/2}}$$

$$a = 464.2 \text{ pm } (= 4.64 \text{ Å})$$

27. The contributions of the four atoms in contact forming the tetrahedral void may be represented at the corners *A*, *B*, *C* and *D*. A plane through *A* and *B* and bisecting the line $\overline{CD}$ is represented as follows:

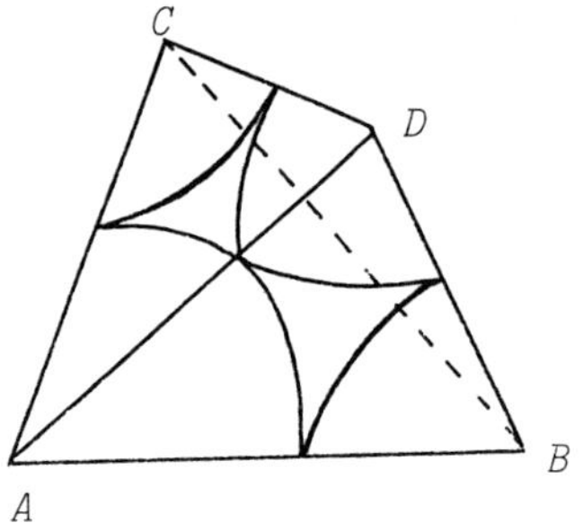

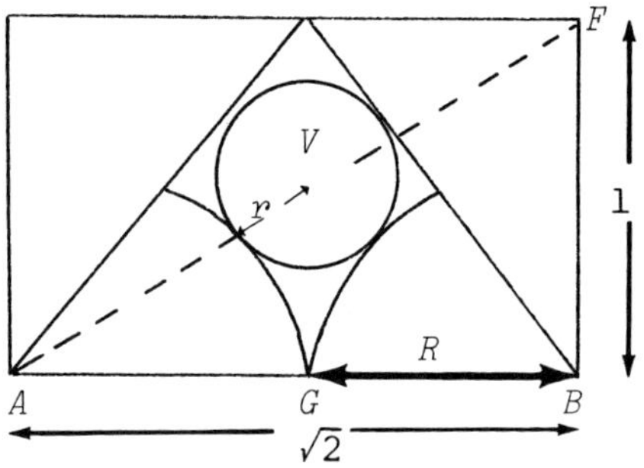

Here the atoms at *A* and *B* are shown by arcs. The right triangle *AGV* is similar to the right triangle *ABF*. Therefore,

$$\frac{AG}{AV} = \frac{AB}{AF} = \sqrt{\frac{2}{3}}$$

Then the maximum radius of the circle representing the void can be no more than *AV* - *R*, that is

$$\frac{AG}{AV} = \frac{R}{R + r} = \sqrt{\frac{2}{3}}$$

$$r = \frac{\sqrt{3} - \sqrt{2}}{2} R = 0.225\ R$$

28. Take a section through an octahedron with sides of unit length:

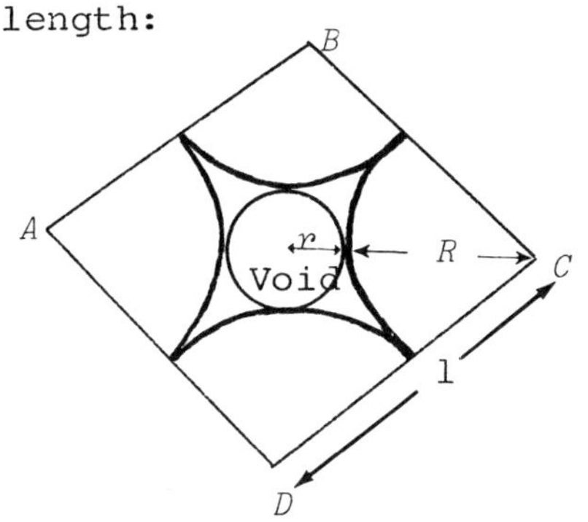

The diagonal AC is $\sqrt{2}$.

From the isosceles right triangle,

$$\frac{AC}{AB} = \frac{\sqrt{2}}{1}$$

If r is the radius of the void then

$$\frac{2R + 2r}{2R} = \frac{\sqrt{2}}{1}$$

$$r = \sqrt{2}R - R = 0.414\ R$$

29. From Eq. 15.37, $\Delta E_c = -\Delta_f H + \Delta_{sub}H + \frac{1}{2}D_o + I - A$

$= (414 + 84 + \frac{1}{2}(192) + 397 - 318)\ \text{kJ mol}^{-1}$

$= 668\ \text{kJ mol}^{-1}$

30. (a) $n\ \lambda = 2\ d \sin\ \theta$

d/pm	323	309	304	284	274
θ	13.81°	14.45°	14.69°	15.75°	16.33°

(b) $$\frac{1}{d^2} = \frac{h^2 + k^2 + l^2}{a^2}$$

$$d = 323 \text{ pm} \quad \frac{1}{(d/\text{pm})^2} = \frac{1}{104\ 300} = \frac{h^2 + k^2 + l^2}{4\ 639\ 700}$$

$$h^2 + k^2 + l^2 = 44.48 \text{ pm}^2$$

$$6^2 + 2^2 + 2^2 = 36 + 4 + 4 = 44$$

$$d = 309 \text{ pm} \quad \frac{1}{(d/\text{pm})^2} = \frac{1}{95\ 481} = \frac{h^2 + k^2 + l^2}{4\ 639\ 700}$$

$$h^2 + k^2 + l^2 = 48.63 \text{ pm}^2$$

$$4^2 + 4^2 + 4^2 = 48$$

$$d = 304 \text{ pm} \quad \frac{1}{(d/\text{pm})^2} = \frac{1}{92\ 400} = \frac{h^2 + k^2 + l^2}{4\ 639\ 700}$$

$$h^2 + k^2 + l^2 = 50.2 \text{ pm}^2$$

$$5^2 + 4^2 + 3^2 = 50$$

Note that other combinations of *hkl* values are possible, as well as different orders of the values given. As higher *hkl* values are used more possibilities exist, making it more difficult to decide on the correct values. Help in this task is provided by knowledge of the lines that are forbidden for the particular crystal type, and of the extinction caused by absorption by different atoms in the crystal layers.

31. $n\lambda = 2\ d \sin\theta \qquad d = \lambda/2 \sin\theta$

$d_{12.95°} = 344$ pm $d_{13.76°} = 324$ pm $d_{14.79°} = 302$ pm

$$\frac{1}{d^2} = \frac{h^2}{a^2} + \frac{k^2}{b^2} + \frac{l^2}{c^2}$$

With the values given,

$$\frac{1}{(d/\text{pm})^2} = \frac{h^2}{822.7^2} + \frac{k^2}{1198.2^2} + \frac{l^2}{644.1^2}$$

$$= \frac{h^2}{676\ 800} + \frac{k^2}{1\ 436\ 000} + \frac{l^2}{414\ 900}$$

With d = 344 pm, agreement is obtained with h = 2, k = 0, and l = 1:

$$\text{LHS} = \frac{1}{344^2} = 8.45 \times 10^{-6}$$

$$\text{RHS} = 5.90 \times 10^{-6} \times 0 \times 2.40 \times 10^{-6} = 8.30 \times 10^{-6}$$

With d = 324 pm, agreement is obtained with h = 0, k = 0, and l = 2;

$$\text{LHS} = \frac{1}{324^2} = 9.52 \times 10^{-6}$$

$$\text{RHS} = 0 + 0 + 9.60 \times 10^{-6} = 9.60 \times 10^{-6}$$

With d = 302 pm, agreement is obtained with h = 0, k = 14, and l = 0

$$\text{LHS} = \frac{1}{302^2} = 10.96 \times 10^{-6}$$

$$\text{RHS} = 0 + 11.1 \times 10^{-6} + 0 = 11.1 \times 10^{-6}$$

In the latter case agreement is obtained with h = 1, k = 0, and l = 2 (RHS = 11.0×10^{-6}) but in view of the crystal type (040) is more likely.

32. (a) From Eq. 15.21

$$F(hkl) = \sum_{j=1}^{N} f_j \exp[2\pi i\,(hx_j + ky_j + lz_j)]$$

$$= f_{Zn}[\exp 2\pi i(0) + \exp 2\pi i(\tfrac{1}{2} + \tfrac{1}{2})$$

$$+ \exp 2\pi i(\tfrac{1}{2} + \tfrac{1}{2}) + \exp 2\pi i(\tfrac{1}{2} + \tfrac{1}{2})] +$$

$$f_S[\exp 2\pi i(\tfrac{1}{4} + \tfrac{1}{4} + \tfrac{1}{4}) + \exp 2\pi i(\tfrac{1}{4} + \tfrac{3}{4} + \tfrac{3}{4}) +$$

$$\exp 2\pi i(\tfrac{3}{4} + \tfrac{1}{4} + \tfrac{3}{4}) + \exp 2\pi i(\tfrac{3}{4} + \tfrac{3}{4} + \tfrac{1}{4})] \quad F\,(hkl)$$

$$= f_{Zn}[1 + 3e^{2\pi i}] + f_S[e^{3\pi i/2} + 3e^{\frac{7\pi i}{2}}]$$

$$= 4\, f_{Zn} - 4\, i\, f_S$$

(a) $$a = \frac{\lambda}{2\sin\theta}(h^2 + k^2 + l^2)^{1/2} = \frac{154.18\ \text{pm}\ (3)^{1/2}}{2(0.247)}$$

$$= 540.5\ \text{pm}$$

33. The number of atoms per unit volume must be calculated

$$N/V = \frac{L\rho}{A}$$

where A is the atomic mass and ρ is the density.

$$\nu_D = \left(\frac{9N}{4\pi V}\right)^{1/3}\left(\frac{1}{c_l^{\,3}} + \frac{2}{c_t^{\,3}}\right)^{1/3}$$

$$= \frac{9}{4\pi}\ \frac{(19.271\ \text{g cm}^{-3})(6.022 \times 10^{23}\ \text{mol}^{-1}}{183.85\ \text{g mol}^{-1}} \times$$

$$\left[\frac{1}{(5.2496 \times 10^5\ \text{cm s}^{-1})^3} + \frac{2}{(2.9092 \times 10^5\ \text{cm s}^{-1})^3}\right]^{1/3}$$

$$\nu_D = 7.986 \times 10^{12}\ \text{s}^{-1}$$

Then $$\Theta_D = \frac{h\nu D}{\mathbf{k}} = \frac{6.6262 \times 10^{-34}\ \text{J s}\ (7.986 \times 10^{12}\ \text{s}^{-1})}{1.3807 \times 10^{-23}\ \text{J K}^{-1}}$$

$$= 383\ \text{K}$$

CHAPTER 16: WORKED SOLUTIONS
THE LIQUID STATE

1. The molar mass of ethanol is 46.07 g mol^{-1}, and since the density of liquid ethanol is 0.790 g cm^{-3} the molar volume is 58.32 cm^3 mol^{-1} = 5.832 x 10^{-5} m^3 mol^{-1}. The internal pressure is therefore

$$P_i = \frac{1.218 \text{ Pa m}^6 \text{ mol}^{-2}}{(5.832 \times 10^{-5})^2 \text{ m}^6 \text{ mol}^{-2}} = 3.58 \times 10^8 \text{ Pa}$$

$$= 3530 \text{ atm}$$

$$E_p = -\frac{1.218 \text{ Pa m}^6 \text{ mol}^{-2}}{5.832 \times 10^{-5} \text{ m}^3 \text{ mol}^{-1}} = -20.9 \text{ kJ mol}^{-1}$$

2. $$P_i = \left(\frac{\partial U}{\partial V}\right)_T = T\left(\frac{\partial P}{\partial T}\right)_V - P \qquad \text{(Eq. 16.2)}$$

$$= (298 \times 6.60 \times 10^6 - 101\ 325) \text{ Pa}$$

$$= 1.97 \times 10^9 \text{ Pa} = 19\ 400 \text{ atm}$$

3. The molar mass is 78.11 g mol^{-1}, and the molar volume is 42.82 cm^3 mol^{-1} = 4.282 x 10^{-5} m^3 mol^{-1}. The internal pressure is therefore

$$P_i = \frac{1.824 \text{ Pa m}^6 \text{ mol}^{-2}}{(4.282 \times 10^{-5})^2 \text{ m}^6 \text{ mol}^{-2}} = 9.95 \times 10^8 \text{ Pa}$$

$$= 9820 \text{ atm}$$

$$E_p = \frac{-1.824 \text{ Pa m}^6 \text{ mol}^{-2}}{4.282 \times 10^{-5} \text{ m}^3 \text{ mol}^{-1}} = -42.6 \text{ kJ mol}^{-1}$$

4. $$P_i = T\left(\frac{\partial P}{\partial T}\right)_V - P \qquad \text{(Eq. 16.2)}$$

$$= (298 \times 1.24 \times 10^6 - 101\ 325) \text{ Pa}$$

$$= 3.69 \times 10^8 \text{ Pa} = 3650 \text{ atm}$$

5. (a) Hg: $P_i = (4.49 \times 10^6 \times 298 - 101\ 325)$

$$= 1.34 \times 10^9 \text{ Pa} = 13\ 200 \text{ atm}$$

(b) *n*-Heptane: $P_i = (8.53 \times 10^5 \times 298 - 101\ 325)$

$= 2.54 \times 10^8$ Pa = 2500 atm

(c) *n*-Octane: $P_i = (1.01 \times 10^6 \times 298 - 101\ 325)$

$= 3.01 \times 10^8$ Pa

= 2970 atm

(d) Diethyl ether: $P_i = (8.06 \times 10^5 \times 298 - 101\ 325)$

$= 2.40 \times 10^8$ Pa = 2370 atm

6. For the vapor

$$P_i = \left(\frac{\partial U}{\partial V}\right)_T = T\left(\frac{\partial P}{\partial T}\right)_V - P \qquad \text{(Eq. 16.2)}$$

$= (298 \times 115 - 10)$ Pa $= 3.43 \times 10^4$ Pa = 0.339 atm

For the liquid

$P_i = (298 \times 1.24 \times 10^6 - 101\ 325)$ Pa

$= 3.69 \times 10^8$ Pa = 3646 atm

7. (a) $P_i = \dfrac{\alpha T}{\kappa} - P$ (from Eq. 16.2)

$$= \frac{1.06 \times 10^{-3} \times 293}{9.08 \times 10^{-10}} - 101\ 325 \text{ Pa}$$

$= 3.42 \times 10^8$ Pa = 3380 atm

(b) Molar mass of acetic acid = 60.0 g mol^{-1}

Molar volume, $V_m = 60.0/1.049 = 57.2$ cm^3 mol^{-1}

$= 5.72 \times 10^{-5}$ m^3 mol^{-1}

$$P_i = \frac{1.78}{(5.72 \times 10^{-5})^2} \text{ Pa} = 5.44 \times 10^8 \text{ Pa}$$

= 5370 atm

8. Since the intermolecular energy is inversely proportional to the sixth power of the intermolecular distance it is inversely proportional

to the square of the volume. The volume has increased by a factor of 10^3 and the energy therefore changes by a factor of 10^{-6}.

9. $$E_p = -\frac{z_A e \mu}{4\pi\varepsilon_0 r^2} \qquad \text{(from Eq. 16.13)}$$

$$= \frac{-2 \times 1.602 \times 10^{-19}\ \text{C} \times 6.18 \times 10^{-30}\ \text{C m}}{4\pi(8.854 \times 10^{-12}\ \text{C}^2\ \text{N}^{-1}\ \text{m}^{-2})(5.0 \times 10^{-10}\ \text{m})^2}$$

$$= -7.12 \times 10^{-20}\ \text{J}$$

$$= -42.9\ \text{kJ mol}^{-1}$$

10. $$E_p = \frac{z_A z_B e^2}{4\pi\varepsilon_0 r} \qquad \text{(from Eq. 16.12)}$$

$$= -\frac{2(1.602 \times 10^{-19}\ \text{C})^2}{4\pi(8.854 \times 10^{-12}\ \text{C}^2\ \text{N}^{-1}\ \text{m}^{-2})(5.0 \times 10^{-10}\ \text{m})}$$

$$= -9.23 \times 10^{-19}\ \text{J} = -555.6\ \text{kJ mol}^{-1}$$

11. $$E_p = \frac{-\alpha(z_A\ )^2}{8\pi\varepsilon_0 r^4} \qquad \text{(Eq. 16.14)}$$

$$= -\frac{(2.0 \times 10^{-30}\ \text{m}^3)(2 \times 1.602 \times 10^{-19}\ \text{C})^2}{8\pi(8.854 \times 10^{-12}\ \text{C}^2\ \text{N}^{-1}\ \text{m}^{-2})(5.0 \times 10^{-10}\ \text{m})}$$

$$= -1.48 \times 10^{-20}\ \text{J} = -8.91\ \text{kJ mol}^{-1}$$

12. $$E_p = -\frac{\mu_A^2 \mu_B^2}{24\pi^2\varepsilon_0^2\ \mathbf{k}T\ r^6} \qquad \text{(from Eq. 16.17)}$$

$$= -\frac{(6.18 \times 10^{-30}\ \text{C m})^4}{24\pi^2(8.854 \times 10^{-12}\ \text{C}^2\ \text{N}^{-1}\ \text{m}^{-2})^2(1.381 \times 10^{-23} \times 298.15\ \text{J})(5.0 \times 10^{-10}\ \text{m})^6}$$

$$= -1.22 \times 10^{-21}\ \text{J}$$

$$= -740\ \text{J mol}^{-1}$$

13. Differentiation of Eq. 16.21 gives

$$\frac{dE_p}{dr} = \frac{6\ A}{r^7} - \frac{12\ B}{r^{13}}$$

This is zero when $r = r_0$ and therefore

$$r_0 = \left(\frac{2\ B}{A}\right)^{1/6}$$

$$= \left(\frac{2 \times 3.42 \times 10^{10}\ \text{J pm}^{12}}{1.34 \times 10^{-5}\ \text{J pm}^6}\right)^{1/6}$$

$$= 414\ \text{pm}$$

The energy at this separation is

$$E_p = \left(-\frac{1.34 \times 10^{-5}}{(414)^6} + \frac{3.42 \times 10^{10}}{(414)^{12}}\right)\ \text{J}$$

$$= \left(-\frac{1.34 \times 10^{-5}}{5.035 \times 10^{15}} + \frac{3.42 \times 10^{10}}{2.535 \times 10^{31}}\right)\text{J}$$

$$= (-2.66 \times 10^{-21} + 1.35 \times 10^{-21})\text{J}$$

$$= -1.31 \times 10^{-21}\ \text{J} = -790\ \text{J mol}^{-1}$$

14. The dipole-dipole energy is given by Eq. 16.17:

$$E_p = -\frac{(2.60 \times 10^{-30})^4}{24\pi^2(8.854 \times 10^{-12})^2(1.381 \times 10^{-23} \times 298.15)} \times 1/(5 \times 10^{-10})^6$$

$$= -3.83 \times 10^{-23}\ \text{J} = 23.0\ \text{J mol}^{-1}$$

The dipole-(induced dipole) energy is given by Eq. 16.18.

$$E_p = -\frac{3.58 \times 10^{-30}(2.60 \times 10^{-30})^2}{2\pi(8.854 \times 10^{-12})(5 \times 10^{-10})^6}$$

$$-2.79 \times 10^{-23}\ \text{J} = -16.8\ \text{J mol}^{-1}$$

The potential energy due to the dispersion forces is given by Eq. 16.19:

$$E_p = -3(6.626 \times 10^{-34} \times 3.22 \times 10^{15}\ \text{J}) \times (3.58 \times 10^{-30}\ \text{m}^3)^2/4(5 \times 10^{-10})^6$$

$$= -1.31 \times 10^{-21}\ \text{J} = -790\ \text{J mol}^{-1}$$

15. $$E_d = -\frac{3h\nu_0\alpha_0^2}{4r^6}$$

$$E_d/\text{J} = -\frac{3 \times 6.626 \times 10^{-34}\ (\nu_0/\text{s}^{-1})(\alpha/\text{m}^3)^2}{4 \times (500 \times 10^{-12})^6}$$

$$= -3.180 \times 10^{22}\ (\nu_0/\text{s}^{-1})(\alpha/\text{m}^3)^2$$

For Ne, $E_d/\text{J} = -3.180 \times 10^{22} \times 5.21 \times 10^{15} \times (0.39 \times 10^{-30})^2$

$$E_d = -2.52 \times 10^{-23}\ \text{J} = -15.2\ \text{J mol}^{-1}$$

For Ar, $E_d/\text{J} = -3.180 \times 10^{22} \times 3.39 \times 10^{15} \times (1.63 \times 10^{-30})^2$

$$E_d = -2.86 \times 10^{-22}\ \text{J} = -172\ \text{J mol}^{-1}$$

For Kr, $E_d/\text{J} = -3.180 \times 10^{22} \times 2.94 \times 10^{15} \times (2.46 \times 10^{-30})^2$

$$E_d = -5.66 \times 10^{-22}\ \text{J} = -341\ \text{J mol}^{-1}$$

16. The values previously calculated (Problem 15) will be multiplied by the factor

$$\left(\frac{500}{r/\text{pm}}\right)^6$$

The values of $-E_d$/J mol^{-1} are then:

	He	Ne	Ar	Kr	Xe
For 500 pm:	4.6	15.2	172	341	850
For r:	376	221	892	1300	2420

If each Ar atom has 12 nearest neighbors the estimated enthalpy of vaporization is

$$(12/2) \times 892 \text{ J mol}^{-1}$$

$$= 5.3 \text{ kJ mol}^{-1}$$

This is not bad agreement considering the simplicity of the model.

17. The interaction energy with $\varepsilon = 1$ is

$$E_p = -\frac{\alpha\mu^2}{4\pi\varepsilon_0 r^6} \qquad \text{(compare Eq. 16.18)}$$

Then for $Ar{\cdot}H_2O$,

$$E_p = -\frac{(1.63 \times 10^{-30} \text{ m}^3)(6.18 \times 10^{-30} \text{ C m})^2}{4\pi(8.854 \times 10^{-12} \text{ C}^2 \text{ N}^{-1} \text{ m}^{-2})(6.0 \times 10^{-10} \text{ m})^6}$$

$$= -1.20 \times 10^{-23} \text{ J} = 7.22 \text{ J mol}^{-1}$$

In $Ag{\cdot}5H_2O$ there will in addition be considerable hydrogen bonding between neighboring water molecules.

18. (a) $$E_p = -\frac{A}{r^6} + \frac{B}{r^n} \qquad (1)$$

Differentiating, with $r = r_0$ at the minimum,

$$\frac{dE_p}{dr} = \frac{6A}{r_0^7} - \frac{nB}{r_0^{n+1}} = 0 \qquad (2)$$

$$E_{min} = -\frac{A}{r_0^6} + \frac{B}{r_0^n} \qquad (3)$$

From (2), $$B = \frac{6Ar_0^{n-6}}{n} \qquad (4)$$

Substituting Eq. 4 in Eq. 3,

$$E_{min} = -\frac{A}{r_0^6} + \frac{6A}{nr_0^6} \qquad (5)$$

$$\text{or } A = \frac{nE_{min}r_0^6}{6 - n} \qquad (6)$$

Insertion of Eq. 6 in Eq. 4,

$$B = \frac{6r_0^n E_{min}}{6 - n} \qquad (7)$$

Then from Eq. 1,

$$\frac{E}{E_{min}} = -\frac{n}{6 - n}\left(\frac{r_0}{r}\right)^6 + \frac{6}{6 - n}\left(\frac{r_0}{r}\right)^n \qquad (8)$$

(b) $$-\frac{A}{(r^*)^6} + \frac{B}{(r^*)^n} = 0 \qquad (9)$$

and thus

$$(r^*)^{n-6} = \frac{B}{A} = \frac{6r_0^{n-6}}{n} \qquad \text{(From Eq. 4)}$$

Therefore

$$\left(\frac{r^*}{r_0}\right)^{n-6} = \frac{6}{n} \qquad (10)$$

(c) If $n = 12$, from Eq. 8

$$E = E_{min}\left[2\left(\frac{r_0}{r}\right)^6 - \left(\frac{r_0}{r}\right)^{12}\right] \qquad (11)$$

If $n = 12, (r_0/r^*)^6 = 2$

and from Eq. 11,

$$E = 4E_{min}\left[\left(\frac{r^*}{r}\right)^6 - \left(\frac{r^*}{r}\right)^{12}\right] \qquad (12)$$

19. (a) $$C_p - C_v = \left[P + \left(\frac{\partial U}{\partial V}\right)_T\right]\left(\frac{\partial V}{\partial T}\right)_P \qquad \text{(Eq. 2.117)}$$

One of the thermodynamic equations of state is

$$\left(\frac{\partial U}{\partial V}\right)_T = -P + T\left(\frac{\partial P}{\partial T}\right)_V \qquad \text{(Eq. 3.128)}$$

and therefore

$$C_p - C_v = \left(\frac{\partial P}{\partial T}\right)_V \left(\frac{\partial V}{\partial T}\right)_P T$$

$$\alpha \equiv \frac{1}{V}\left(\frac{\partial V}{\partial T}\right)_P \quad \text{and} \quad \kappa \equiv -\frac{1}{V}\left(\frac{\partial V}{\partial P}\right)_T$$

$$\frac{\alpha}{\kappa} = -\left(\frac{\partial V}{\partial T}\right)_P \left(\frac{\partial V}{\partial P}\right)_T = \left(\frac{\partial P}{\partial T}\right)_V \qquad \text{(see Appendix C)}$$

Thus

$$C_p - C_v = \frac{\alpha^2 VT}{\kappa}$$

(b) For CCl_4, from the data given,

$$C_{p,m} - C_{v,m} = (1.24 \times 10^{-3}\ K^{-1})^2 \times \frac{(97 \times 10^{-6}\ m^3\ mol^{-1})(298.15\ K)}{(10.6 \times 10^{-5}/101\ 325)\ Pa^{-1}}$$

$$= 42.5\ J\ K^{-1}\ mol^{-1}$$

$$C_{p,m} = 89.5 + 42.5 = 132.0\ J\ K^{-1}\ mol^{-1}$$

(c) For acetic acid,

$$V_m = 60.05\ g\ mol^{-1}/1.049\ g\ cm^{-3}$$

$$= 57.2 \times 10^{-6}\ m^3\ mol^{-1}$$

$$C_{p,m} - C_{v,m} = (1.06 \times 10^{-3}\ K^{-1})^2 \times \frac{(57.2 \times 10^{-6}\ m^3\ mol^{-1})(293\ K)}{9.08 \times 10^{-10}\ Pa^{-1}}$$

$$= 20.7\ J\ K^{-1}\ mol^{-1}$$

CHAPTER 17: WORKED SOLUTIONS
SURFACE CHEMISTRY AND COLLOIDS

1. (a) $$0.5 = \frac{K \times 1\ \text{atm}}{1 + K \times 1\ \text{atm}}$$

$$0.5 + 0.5\ K\ \text{atm} = K\ \text{atm}$$

$$K/\text{atm}^{-1} = 1$$

(b) $$0.75 = \frac{P \times 1\ \text{atm}^{-1}}{1 + P \times 1\ \text{atm}^{-1}};\quad P = 3\ \text{atm}$$

$$0.90 = \frac{P\ \text{atm}^{-1}}{1 + P\ \text{atm}^{-1}};\quad P = 9\ \text{atm}$$

$$0.99 = \frac{P\ \text{atm}^{-1}}{1 + P\ \text{atm}^{-1}};\quad P = 99\ \text{atm}$$

$$0.999 = \frac{P\ \text{atm}^{-1}}{1 + P\ \text{atm}^{-1}};\quad P = 999\ \text{atm}$$

(c) $$\theta = \frac{0.1}{1 + 0.1} = 0.91$$

$$\theta = \frac{0.5}{1 + 0.5} = 0.33$$

$$\theta = \frac{1000}{1 + 1000} = 0.999$$

2. The Langmuir isotherm in terms of pressure P is, from Eq. 17.6,

$$\theta = \frac{KP}{1 + KP}$$

$\theta = V_a/V_a^{\circ}$ and therefore

$$\frac{V_a}{V_a^{\circ}} = \frac{KP}{1 + KP}$$

which rearranges to

$$\frac{P}{V_a} = \frac{1 + KP}{V_a^{\circ}K} = \frac{1}{V_a^{\circ}K} + \frac{P}{V_a^{\circ}}$$

A plot of P/V_a against P is therefore linear; the slope is $1/V_a^{\circ}$ and the intercept on the P/V_a axis is $1/V_a^{\circ}K$. The quantities V_a° and K can thus be obtained separately.

3. The amount x adsorbed is proportional to θ and therefore

$$x = \frac{aK[\mathrm{A}]}{1 + K[\mathrm{A}]}$$

To convert atmospheres to concentrations:

$$\frac{n}{V} = \frac{P}{RT} = \frac{P/\mathrm{atm}}{0.082\ 05 \times 293.15\ \mathrm{dm^3\ mol^{-1}}}$$

To convert amount of gas adsorbed to mol:

$$n = \frac{PV}{RT} = \frac{1\,(\mathrm{atm})\,V}{0.082\ 05 \times 293.15\ \mathrm{atm\ dm^3\ mol^{-1}}}$$

$$= \frac{V/\mathrm{mm^3}}{10^6 \times 0.082\ 05 \times 293.15}\ \mathrm{mol}$$

Thus the table becomes

Concentration / $\mathrm{mol\ dm^{-3}}$	0.116	0.166	0.249	0.391	0.711	1.39
Amount adsorbed / 10^{-7} mol	4.99	6.28	7.90	9.94	11.7	13.7

(a) A linear plot may be obtained by plotting $1/x$ against $1/[\mathrm{A}]$:

$$\frac{1}{x} = \frac{1}{aK[\mathrm{A}]} + \frac{1}{a}$$

$[A]^{-1}/dm^3\ mol^{-1}$	8.62	6.02	4.02	2.56	1.41	0.719
$x^{-1}/10^6\ mol^{-1}$	2.00	1.59	1.27	1.01	0.85	0.73

From a plot of x^{-1} against $[A]^{-1}$,

$$a = 1.67 \times 10^{-6}\ mol$$

$$K = 3.72\ dm^3\ mol^{-1}$$

(b) Complete coverage corresponds to

$1.67 \times 10^{-6}\ mol = 1.006 \times 10^{18}$ molecules

The surface area was thus about

$$10^3\ cm^2 = 10^{-1}\ m^2$$

4. (a) The rate and the rate constant are both increased by a factor of 10:

$v = 1.5 \times 10^{-3}\ mol\ dm^{-3}\ s^{-1}$; $k = 2.0 \times 10^{-2}\ s^{-1}$

(b) The rate of conversion ($mol\ s^{-1}$) remains the same, but since the volume is increased by a factor of 10 the rate is reduced by a factor of 10, as is the rate constant:

$v = 1.5 \times 10^{-5}\ mol\ dm^{-3}\ s^{-1}$; $k = 2.0 \times 10^{-4}\ s^{-1}$

(c) Increasing the radius by a factor of 10 increases the surface area by a factor of 100 and the volume by a factor of 1000. The rate and the rate constant are thus reduced by a factor of 10:

$v = 1.5 \times 10^{-5}\ mol\ dm^{-3}\ s^{-1}$; $k = 2.0 \times 10^{-4}\ s^{-1}$

(d) Since k is proportional to S and inversely proportional to V, the constant $k' = kV/S$ is independent of V and S.

(e) Its SI unit is m s^{-1}.

5. The arguments are the same as in Problem 4 and thus

(a) $v = 2.5 \times 10^{-2}$ mol dm^{-3} s^{-1};

$k = 2.5 \times 10^{-2}$ mol dm^{-3} s^{-1}

(b) $v = 2.5 \times 10^{-4}$ mol dm^{-3} s^{-1};

$k = 2.5 \times 10^{-4}$ mol dm^{-3} s^{-1}

(c) $v = 2.5 \times 10^{-4}$ mol dm^{-3} s^{-1};

$k = 2.5 \times 10^{-4}$ mol dm^{-3} s^{-1}

$k' = kV/S$; mol m^{-2} s^{-1}

6.

	$2\,NH_3$	=	N_2	+ $3\,H_2$
Initial concentrations:	a_0		0	0
Concentrations after time t	$a_0 - \frac{2x}{3}$		$\frac{x}{3}$	x

$$\frac{dx}{dt} = k(a_0 - \frac{2x}{3})/x$$

$$= \frac{3\,ka_0 - 2\,kx}{3\,x} = \frac{ka_0}{x} - \frac{2k}{3}$$

7. From Eq. 17.14, fraction of bare surface is

$$1 - \theta = \frac{1}{1 + K^{1/2}[H_2]^{1/2}}$$

Rate of H atom formation is thus

$$v = k(1 - \theta)[H_2]$$

$$= \frac{k[H_2]}{1 + K^{1/2}[H_2]^{1/2}}$$

Kinetics are one-half order when $K^{1/2}[H_2]^{1/2} >> 1$;

i.e. when the surface is fully covered:

$$v = \frac{k}{K^{1/2}}[H_2]^{1/2}$$

8. $k \propto e^{-E/RT}$, $K \propto e^{-\Delta H_A/RT}$ and $K_i \propto e^{-\Delta H_I/RT}$

(a) $v = kK[A] \propto e^{-(E+\Delta H_A)RT}$

(b) $v = k \propto e^{-E/RT}$

(c) $v = \frac{kK}{K_i}\frac{[A]}{[I]} \propto e^{-(E+\Delta H_A - \Delta H_I)/RT}$

9.

	A	$\longrightarrow$	Y	+	Z
Initially	a_0		0		0
At time t	$a_0 - z$		z		z

$$\frac{dz}{dt} = k(a_0 - z) + k_s z$$

$$\frac{dz}{ka_0 + (k_s - k)z} = dt$$

Put $ka_0 - (k_s - k)z = y$; $dy = -(k_s - k)dz$

$$-\frac{1}{(k_s - k)}\int\frac{dy}{y} = \int dt$$

$$-\frac{1}{(k_s - k)} \ln [ka_0 - k)z] = t + I$$

The boundary condition is that $z = 0$ when $t = 0$:

$$I = -\frac{1}{(k_s - k)} \ln ka_0$$

$$t = \frac{1}{(k_s - k)} \ln \frac{ka_0}{ka_0 - (k_s - k)z}$$

$$\frac{ka_0 - (k_s - k)z}{ka_0} = e^{-(k_s - k)t}$$

$$z = \frac{ka_0}{k_s - k}\left[1 - e^{-(k_s - k)t}\right]$$

10. (a) The general rate equation is Eq. 17.34. At low pressures the surface is sparsely covered and

$$v = kK[A]$$

At high pressures it is fully covered and

$$v = k$$

(b) Reaction occurs on certain surface sites on which N_2 is not adsorbed. Rate equation is Eq. 17.36.

(c) A Langmuir-Rideal mechanism. General equation is Eq. 17.44; $K_{H_2}[H_2]$ is small and $K_{O_2}[O_2]$ large so that

$$v = k[H_2]$$

(d) The mechanism is

$$p - H_2 + \text{-}\overset{|}{S}\text{-}\overset{|}{S}\text{-} \rightleftharpoons \text{-}\overset{\overset{H}{|}}{S}\text{-}\overset{\overset{H}{|}}{S}\text{-} \rightleftharpoons \text{-}\overset{|}{S}\text{-}\overset{|}{S}\text{-} + o\text{-}H_2$$

with the surface fully covered. Rate is

$$v = k[H_2](1 - \theta)^2$$

where $1 - \theta$ is given by Eq. 17.15; thus

$$v = \frac{k}{K}$$

11. According to Eq. 17.54 the capillary rise is given by $h = \dfrac{2\gamma}{\rho g r}$

(a) If $r = 10^{-3}$ m,

$$h = \frac{2 \times 7.27 \times 10^{-2}\ (\text{N m}^{-1})}{998(\text{kg m}^{-3})\,9.81(\text{m s}^{-2})\,10^{-3}(\text{m})}$$

$$= 1.49 \times 10^{-2}\ \text{m} = 1.49\ \text{cm}$$

(b) If $r = 10^{-5}$ m,

$$h = 1.49\ \text{m}$$

12. From Eq. 17.55,

$$h = \frac{2\gamma\cos\theta}{\rho g r}$$

$$= \frac{2 \times 0.47\ (\text{N m}^{-1})(-0.766)}{1.36 \times 10^{4}(\text{kg m}^{-3})\,9.81(\text{m s}^{-2}) \times 5 \times 10^{-4}(\text{m})}$$

$$= -10.8 \times 10^{-3}\ \text{m} = -10.8\ \text{mm}$$

13. Volume of droplet $= \dfrac{10^{-12}}{0.998} = 1.002 \times 10^{-12}\ \text{cm}^3$

$$= 1.002 \times 10^{-18}\ \text{m}^3 = \frac{4}{3}\pi r^3$$

$$r^3 = \frac{3 \times 1.002 \times 10^{-18}}{4\pi}$$

$$= 2.392 \times 10^{-19}\ \text{m}^3$$

$$r = 6.21 \times 10^{-7}\ \text{m}$$

$$\ln\frac{P}{P_0} = \frac{2\gamma M}{\rho r R T}$$

$$= \frac{2 \times 7.27 \times 10^{-2}(\text{N m}^{-1}) \times 18.02 \times 10^{-3}(\text{kg mol}^{-1})}{0.998 \times 10^{-3}(\text{kg m}^{-3}) \times 6.21 \times 10^{-7}(\text{m}) \times 8.314} \times \frac{1}{298.15(\text{J mol}^{-1})}$$

$= 0.0017$

$\frac{P}{P_0} = 1.0017$

14. From Eq. 17.54 the height is $2\gamma/r\rho g$ and the difference in heights is

$$\Delta h = \frac{2\gamma}{\rho g}\left(\frac{1}{r_1} - \frac{1}{r_2}\right)$$

Thus

$$0.022 \text{ m} = \frac{2\gamma(2000 - 1000) \text{ m}^{-1}}{0.80 \times 10^3 (\text{kg m}^{-3}) 9.81 (\text{m s}^{-2})}$$

$$= 0.255 \text{ (kg}^{-1} \text{ m s}^2)\gamma$$

$$\gamma = 0.086 \text{ kg s}^{-2} \equiv 0.086 \text{ N m}^{-1}$$

15. The rise is proportional to γ/d and is therefore the same in the second liquid, i.e. is 1.5 cm.

16. No; the water does not flow over the edge. The meniscus will rise to the top of the tube and then the radius of curvature will decrease until the capillary pressure just balances the pressure of the column of liquid; equilibrium is then established. This will occur when the radius of curvature at the surface is half what it is in a longer tube.

17. The equation that applies is an extension of Eq. 17.55:

$$\gamma = \frac{rh\Delta\rho g}{2 \cos \theta}$$

where $\Delta\rho$ is the difference between the two densities.

Then

$$\gamma = (1.00 - 0.80) \times 10^3 \text{ kg m}^{-3} \times 9.81 \text{ m s}^{-2} \times \frac{0.040 \text{ m} \times 0.5 \times 10^{-4} \text{ m}}{2 \times 0.76}$$

$$= 2.58 \times 10^{-3} \text{ N m}^{-1} \qquad (\text{N} \equiv \text{kg m s}^{-2})$$

18. At the three lower pressures πA is constant within the experimental error and has a value of

$$1.11 \times 10^4 \text{ N m kg}^{-1} \quad (= \text{J kg}^{-1})$$

πA is equal to RT and therefore

$$\pi A = 8.314 \times 288.15 = 2396 \text{ J mol}^{-1}$$

The molar mass is thus

$$\frac{2396}{11\ 100} = 0.216 \text{ kg mol}^{-1} = 216 \text{ g mol}^{-1}$$

At the highest pressure the area is 5.7 $\text{cm}^2\ \mu\text{g}^{-1}$.

Since 216 g = 6.022×10^{23} molecules

$$1\ \mu\text{g} = \frac{6.022 \times 10^{23} \times 10^{-6}}{216}$$

$$= 2.79 \times 10^{15} \text{ molecules}$$

Thus 1 molecule occupies

$$\frac{5.7}{2.79 \times 10^{15}} = 2.04 \times 10^{-15} \text{ cm}^2 = 0.204 \text{ nm}^2$$

19. The relative molar mass of 1-hexadecanol is 242.43, and 52.0 μg therefore contains 2.14×10^{-7} mol $= 1.29 \times 10^{17}$ molecules.

Length	Area	Force	Surface Pressure	Area per molecule
cm	cm^2	10^{-5} N	10^{-4} N m^{-1}	nm^2
20.9	292.6	4.14	3.00	0.227
20.3	284.2	8.56	6.20	0.220
20.1	281.4	26.2	19.0	0.218
19.6	274.4	69.0	50.0	0.212
19.1	267.4	108	78.3	0.207
18.6	260.4	234	169.6	0.202
18.3	256.2	323	234.1	0.198
18.1	253.4	394	285.5	0.196
17.8	249.2	531	384.8	0.193

From the graph, area for the fully-compressed layer $= 0.19$ nm^2. See page 295.

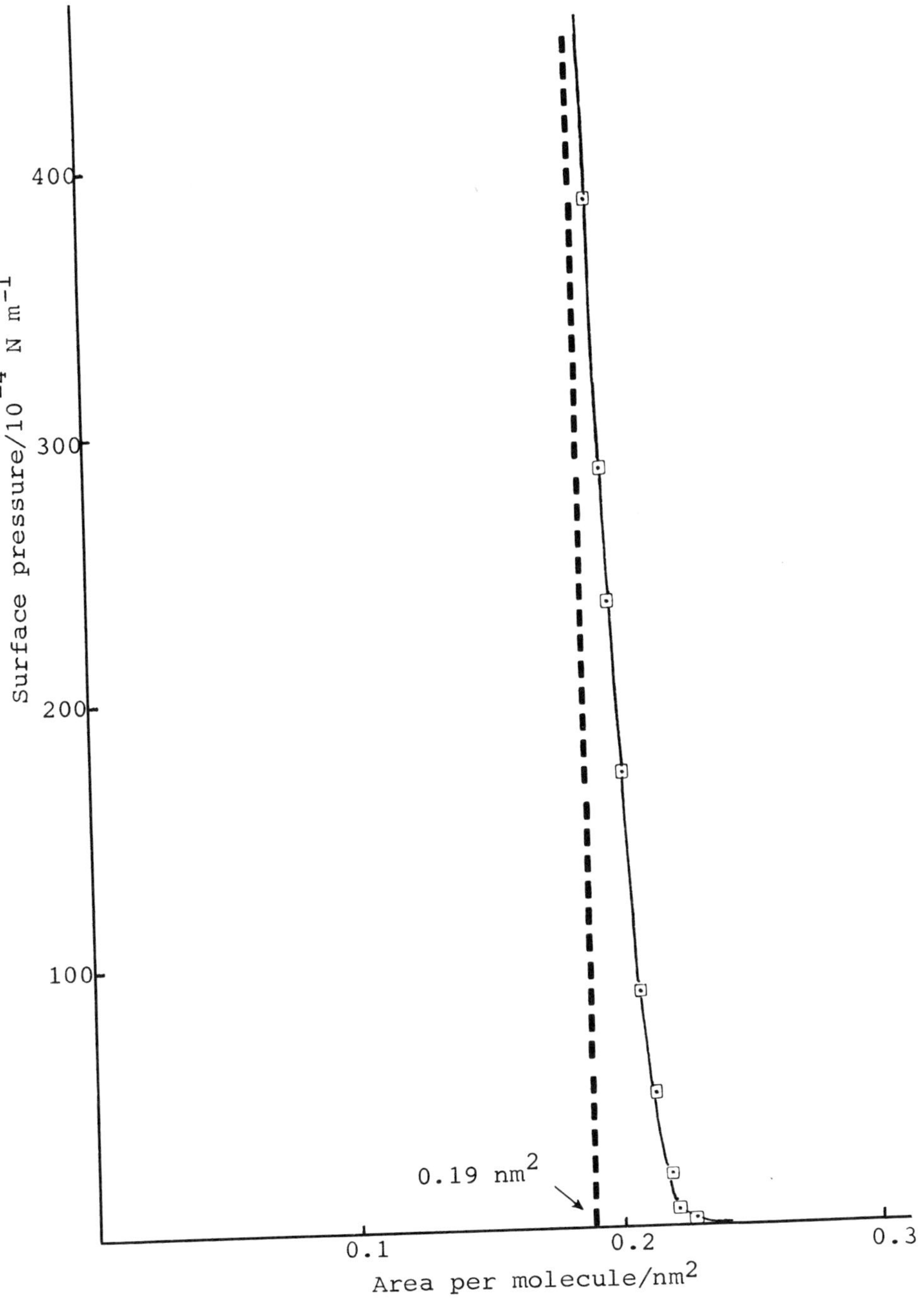
400
300
200
100
Surface pressure/10^{-4} N m^{-1}
0.19 nm^2
0.1
0.2
0.3
Area per molecule/nm^2

20. (a) From Eq. 17.7

$$1 - \theta = \frac{1}{1 + KP} \tag{1}$$

and therefore

$$\theta = KP(1 - \theta) \tag{2}$$

Then

$$\ln \frac{\theta}{P} = \ln K + \ln (1 - \theta) \tag{3}$$

$$\approx \ln K - \theta \text{ if } \theta << 1 \tag{4}$$

A plot of ln (θ/P) against ln θ is thus linear with a slope of -1.

(b) Since $\theta = V_a/V_a^\circ$, eq. (4) can be written as

$$\ln (V_a/P) - \ln V_a^\circ \approx \ln K - V_a/V_a^\circ \tag{5}$$

A plot of ln (V_a/P) against V_a thus has a slope of $-1/V_a^\circ$.

21. (a) The BET isotherm can be tested in a number of ways, for example by plotting $P/V_a(P^\circ - P)$ against P.

(b) If $P_0 >> P$, the isotherm becomes

$$\frac{P}{V_a} = \frac{1}{V_0^\circ K} + \frac{P}{V_a^\circ}$$

The fraction covered $\theta = V_a/V_a^\circ$ and therefore

$$\frac{P}{\theta} = \frac{1}{K} + P$$

or

$$\theta = \frac{KP}{1 + KP}$$

which is the Langmuir isotherm.

22. The process is

$$A_2 + 2\ S \rightleftharpoons 2\ S\text{-}H$$

and

$$K_c = \frac{c_a^{\ 2}}{c_g c_s^{\ 2}} = \frac{N_a^{\ 2}}{(N_g/V)\ N_s^{\ 2}}$$

$$= \left(\frac{\theta}{1 - \theta}\right)^2 \frac{1}{c_g}$$

In terms of partition functions

$$K_c = \frac{q_a^{\ 2}}{g_g g_s^{\ 2}}\ \mathrm{e}^{-\Delta E_0/RT}$$

$$= \frac{h^3 b_a^{\ 2}}{(2\pi m\mathbf{k}T)^{3/2} b_g}\ \mathrm{e}^{-\Delta E_0/RT} \quad (\text{if } q_s = 1)$$

Therefore

$$\frac{\theta}{1 - \theta} = c_g^{\ 1/2}\ \frac{h^{3/2}\ b_a^{\ 1/2}}{(2\pi m\mathbf{k}T)^{3/4} b_g^{\ 1/2}}\ \mathrm{e}^{-\Delta E_0/2\ RT}$$

23. The adsorption centers need no longer be regarded as reactants; the equilibrium is between gas molecules and molecules forming the two-dimensional layer;

$$K_c = \frac{c_a}{c_g} = \frac{N_a/S}{N_g/V} = \frac{\dfrac{(2\pi m\mathbf{k}T)}{h^2}\ b_a}{\dfrac{(2\pi m\mathbf{k}T)^{3/2}}{h^3} b_g}\ \mathrm{e}^{-\Delta E_0/RT}$$

Therefore

$$c_a = c_g\ \frac{h}{(2\pi m\mathbf{k}T)^{1/2}}\ \frac{b_a}{b_g}\ \mathrm{e}^{-\Delta E_0/RT}$$

1. Area of cross-section of aorta

$$= \pi(9 \times 10^{-3})^2 = 2.54 \times 10^{-4}\ m^2$$

Volume rate of flow $= 2.54 \times 10^{-4} \times 0.33\ m^3\ s^{-1}$

$= 8.40 \times 10^{-5}\ m^3\ s^{-1}$

From Pouiseuille's equation (Eq. 18.10)

$$\Delta P = \frac{8\eta l}{\pi R^4} \cdot \frac{dV}{dt}$$

$$= \frac{8 \times 4 \times 10^{-3}\ (N\ s\ m^{-2})\ 0.5\ (m) \times 8.40 \times 10^{-5}\ m^3\ s^{-1}}{\pi(9 \times 10^{-3})^4\ m^4}$$

$= 65.2$ Pa $= 0.49$ mmHg

2. From Pouiseuille's equation (Eq. 18.10)

$$\frac{dV}{dt} = \frac{\pi R^4 \Delta P}{8\eta l} = \frac{\pi(2 \times 10^{-6})^4 \times 20 \times 133.3}{8 \times 4 \times 10^{-3} \times 10^{-3}}$$

$= 4.19 \times 10^{-15}\ m^3\ s^{-1}$ (b)

(a) Area of cross-section $= \pi(2 \times 10^{-6})^2$

$= 1.26 \times 10^{-11}\ m^2$

Linear rate of flow $= 4.19 \times 10^{-15}/1.26 \times 10^{-11}$

$= 3.33 \times 10^{-4} = 0.33$ mm s^{-1}

(c) Number of capillaries $= \dfrac{8.40 \times 10^{-5}}{4.19 \times 10^{-15}} = 2 \times 10^{10}$

3. From the ideal gas law

$$\frac{n}{V} = \frac{P}{RT} = \frac{101\ 325}{8.314 \times 298.15} = 40.88\ mol\ m^{-3}$$

$$\frac{N}{V} = 40.88 \times 6.022 \times 10^{23}\ m^{-3} = 2.462 \times 10^{25}\ m^{-3}$$

$$\bar{u} = \left(\frac{8\mathbf{k}T}{\pi m}\right)^{1/2}$$

$$m = 28.05/6.022 \times 10^{23} = 4.658 \times 10^{-23}\ g$$

$$= 4.658 \times 10^{-26} \text{ kg}$$

$$\bar{u} = \left(\frac{8 \times 1.381 \times 10^{-23} \times 298.15}{\pi \times 4.658 \times 10^{-26}}\right)^{1/2}$$

$$= 474.4 \text{ m s}^{-1}$$

(a) From Eq. 18.15

$$d = \left(\frac{mu}{2\sqrt{2}\pi\eta}\right)^{1/2}$$

$$= \left(\frac{4.658 \times 10^{-26} \times 474.4}{2\sqrt{2}\pi \times 9.33 \times 10^{-6}}\right)^{1/2}$$

$$= 5.16 \times 10^{-10} \text{ m} = 0.516 \text{ nm}$$

(b) Mean free path, $\lambda = \dfrac{V}{\sqrt{2}\pi d_A^2 N_A}$ (Eq. 1.68)

$$= \frac{1}{\sqrt{2}\pi(5.16 \times 10^{-10})^2 \times 2.462 \times 10^{25}}$$

$$= 3.43 \times 10^{-8} \text{ m}$$

$$= 34.3 \text{ nm}$$

(c) $Z_A = \dfrac{\bar{u}}{\lambda} = \dfrac{474.4}{3.43 \times 10^{-8}} = 1.38 \times 10^{10} \text{ s}^{-1}$

(d) From Eqs. 1.59 and 1.61,

$$Z_{AA} = \frac{1}{2} Z_A N/V$$

$$= \frac{1.38 \times 10^{10} \times 2.462 \times 10^{25}}{2}$$

$$= 1.70 \times 10^{35} \text{ m}^{-3} \text{ s}^{-1}$$

4. (a) At 0°C

$$e^{-E/RT} = e^{-10\,900/8.314 \times 273.15} = 8.232 \times 10^{-3}$$

At 40.0°C

$$e^{-E/RT} = e^{-10\,900/8.314 \times 313.15} = 15.20 \times 10^{-3}$$

$$\text{Viscosity at } 40.0^\circ\text{C} = 1.33 \times 10^{-3} \times \frac{8.232}{15.20}\ \text{kg m}^{-1}\ \text{s}^{-1}$$

$$= 7.20 \times 10^{-4}\ \text{kg m}^{-1}\ \text{s}^{-1}$$

(b) At 20°C

$$e^{-E/RT} = e^{-18\ 000/8.314 \times 293.15} = 6.203 \times 10^{-4}$$

At 40°C

$$e^{-E/RT} = e^{-18\ 000/8.314 \times 313.15} = 9.941 \times 10^{-4}$$

$$\text{Viscosity at } 40.0^\circ\text{C} = 1.002 \times 10^{-3} \times \frac{6.203}{9.941}\ \text{kg m}^{-1}\ \text{s}^{-1}$$

$$= 6.25 \times 10^{-4}\ \text{kg m}^{-1}\ \text{s}^{-1}$$

(c) $$\log_{10}\left(\frac{\eta_{20^\circ}}{\eta_{40^\circ}}\right) = \frac{(1.370\ 23 \times 20) + 8.36 \times 10^{-4}}{149}$$

$$= 0.186$$

$$\eta_{40^\circ} = \eta_{20^\circ} \times 10^{-0.186} = 0.651 \times \eta_{20^\circ}$$

$$= 0.651 \times 1.002 \times 10^{-3}\ \text{kg m}^{-1}\ \text{s}^{-1}$$

$$= 6.53 \times 10^{-4}\ \text{kg m}^{-1}\ \text{s}^{-1}$$

5. (a) $$[\eta] = \frac{0.05}{5.90} \cdot \frac{1}{0.1\ \text{g dm}^{-3}}$$

$$= 0.084\ \text{dm}^3\ \text{g}^{-1} = 0.084\ \text{m}^3\ \text{kg}^{-1}$$

(b) $$[\eta] = \frac{0.15}{5.90} \cdot \frac{1}{0.1\ \text{g dm}^{-3}}$$

$$= 0.254\ \text{dm}^3\ \text{g}^{-1} = 0.254\ \text{m}^3\ \text{kg}^{-1}$$

(c) $$[\eta] = \frac{0.37}{5.90} \cdot \frac{1}{0.1\ \text{g dm}^{-3}} = 0.627\ \text{m}^3\ \text{kg}^{-1}$$

6. Taking logarithms:

(a) $\log_{10}[M] = 4.30 \quad \log_{10}[\eta] = -1.075$

(c) $\log_{10}[M] = 4.60 \quad \log_{10}[\eta] = -0.202$

(b) $\log_{10}[\eta] = -0.595$

If the Mark-Houwink equation applies there is a linear relationship between $\log_{10}[\eta]$ and $\log_{10}[M]$.

$$\frac{-0.595 + 1.075}{-0.202 + 1.075} = 0.55$$

If x = relative molar mass for (b),

$$\log_{10}x = 0.55 \times (4.60 - 4.30) + 4.30 = 4.465$$

$$x = 29\ 200$$

7. The mass of the helium atom is

$$m = 4.0026/6.022 \times 10^{23} = 6.647 \times 10^{-24}\ \text{g}$$

$$= 6.647 \times 10^{-27}\ \text{kg}$$

(a) From Eq. 18.16,

$$\eta = \frac{(6.647 \times 10^{-27} \times 1.381 \times 10^{-23} \times 273.15)^{1/2}}{\pi^{3/2}(0.225 \times 10^{-9})^2}$$

$$= 1.78 \times 10^{-5}\ \text{kg m}^{-1}\ \text{s}^{-1}$$

(b) $$\rho = \frac{mN}{V} \quad \frac{mP}{\mathbf{k}T} = \frac{6.647 \times 10^{-27} \times 1.013\ 25 \times 10^5}{1.381 \times 10^{-23} \times 273.15}$$

$$= 0.1785\ \text{kg m}^{-3}$$

$$D = \frac{\eta}{\rho} = \frac{1.78 \times 10^{-5}}{0.1785} = 9.97 \times 10^{-5}\ \text{m}^2\ \text{s}^{-1}$$

(c) $$\bar{u} = (8\mathbf{k}T/\pi m)^{1/2} = \left(\frac{8 \times 1.381 \times 10^{-23} \times 273.15}{\pi \times 6.647 \times 10^{-27}}\right)^{1/2}$$

$$= 1202\ \text{m s}^{-1}$$

(d) $V/\sqrt{2}\ \pi d^2 N = \lambda$

$$\frac{V}{N} = \frac{1.381 \times 10^{-23} \times 273.15}{1.013\ 25 \times 10^5} = 3.7229 \times 10^{-26}$$

$$\lambda = 3.7229 \times 10^{-26}/\sqrt{2}\pi(0.225 \times 10^{-9})^2$$

$$= 1.655 \times 10^{-7}\ \text{m}$$

(e) $Z_A = \bar{u}/\lambda = 1202/1.655 \times 10^{-7} = 7.263 \times 10^9\ \text{s}^{-1}$

(f) $Z_{AA} = \frac{1}{2} Z_A N/V = \dfrac{7.263 \times 10^9}{3.7229 \times 10^{-26} \times 2}$

$$= 9.75 \times 10^{34}\ \text{m}^{-3}\ \text{s}^{-1}$$

8. From Eq. 18.57,

$\overline{x^2} = 2tD = 2 \times 10 \times 1.005 \times 10^{-4}\ \text{m}^2$

$\left(\overline{x^2}\right)^{1/2} = 0.045\ \text{m} = 4.5\ \text{cm}$

9. $t = 100 \times 26 \times 60 \times 60 = 8.64 \times 10^6\ \text{s}$

(a) glucose: $\overline{x^2} = 2 \times 6.8 \times 10^{-10} \times 8.64 \times 10^6$

$= 0.0118\ \text{m}^2$

$\sqrt{\overline{x^2}} = .108\ \text{m} = 10.8\ \text{cm}$

(b) Tobacco mosaic virus: $\overline{x^2} = 2 \times 5.3 \times 10^{-12} \times 8.64 \times 10^6$

$= 9.12 \times 10^{-5}\ \text{m}$

$\sqrt{\overline{x^2}} = 9.55 \times 10^{-3}\ \text{m}$

$= 0.96\ \text{cm}$

10. $D = \dfrac{8.314 \times 298.15}{(964\ 87)^2 \times 2} \lambda° = 133 \times 10^{-7}\ \lambda°\ \text{cm}^2\ \text{s}^{-1}$

For Cu^{2+}: $D_+ = 1.33 \times 10^{-7} \times 56.6 = 0.755 \times 10^{-5}\ \text{cm}^2\ \text{s}^{-1}$

For SO_4^{2-}: $D_- = 1.33 \times 10^{-7} \times 80.0 = 1.065 \times 10^{-5}$ cm^2 s^{-1}

$$D = \frac{2 \times 0.755 \times 10^{-5} \times 1.065 \times 10^{-5}}{(0.755 + 1.065) \times 10^{-5}} \text{ (from Eq. 18.74)}$$

$$= 8.9 \times 10^{-6} \text{ cm}^2 \text{ s}^{-1}$$

11. From Eq. 18.12

$$D = \frac{RT}{QL} u_e$$

Charge on 1 mol of a univalent ion, QL

$$= 96\ 500 \text{ C mol}^{-1}$$

$$D = \frac{(8.314 \text{ J K}^{-1} \text{ mol}^{-1})(298.15 \text{ K})}{96\ 500 \text{ C mol}^{-1}} u_e$$

$$= (0.0256 \text{ V}) u_e \text{ since J = C V}$$

For Na^+, with $u_e = 5.19 \times 10^{-4}$ cm^2 V^{-1} s^{-1},

$$D = 0.0256 \times 5.19 \times 10^{-4} = 1.33 \times 10^{-5} \text{ cm}^2 \text{ s}^{-1}$$

For CH_3COO^-, with $u_e = 4.24 \times 10^{-4}$ cm^2 V^{-1} s^{-1},

$$D = 0.0256 \times 4.24 \times 10^{-4} = 1.09 \times 10^{-5} \text{ cm}^2 \text{ s}^{-1}$$

For the electrolyte, using Eq. 18.74,

$$D = \frac{2 \times 1.33 \times 10^{-5} \times 1.09 \times 10^{-5}}{1.33 \times 10^{-5} + 1.09 \times 10^{-5}} \text{ cm}^2 \text{ s}^{-1}$$

$$= 1.20 \times 10^{-5} \text{ cm}^2 \text{ s}^{-1}$$

12. From Stokes's law (Eq. 18.77),

$$r = \frac{1.381 \times 10^{-23} \times 293.15}{6 \times 3.1426 \times 1.002 \times 10^{-3} \times 6.3 \times 10^{-11}}$$

$$= 3.40 \times 10^{-9} \text{ m} = 3.4 \text{ nm}$$

$$\text{Volume} = \frac{4}{3}\pi(3.40 \times 10^{-9})^3 = 1.65 \times 10^{-25} \text{ m}^3$$

$$\text{Molecular mass} = 1.65 \times 10^{-25} \times \frac{1}{0.75} \times 10^{6}$$
$$= 2.20 \times 10^{-19}\ \text{g}$$

$$\text{Molar mass} = 2.20 \times 10^{-19} \times 6.022 \times 10^{23}$$
$$= 132\ 000\ \text{g mol}^{-1}$$

13. $t = \overline{x^2}/2D$

$$= \frac{(10 \times 10^{-6})^2}{2 \times 8.2 \times 10^{-11}} = 0.61\ \text{s}$$

14. Radius r of particle = 1.5×10^{-7} m

From Stokes's law (Eq. 18.77)

$$D = \frac{\mathbf{k}T}{6\pi\eta r} = \frac{1.381 \times 10^{-23} \times 293.15\ \text{J}}{6\pi \times 1.002 \times 10^{-3} \times 1.5 \times 10^{-7}\ \text{kg s}^{-1}}$$
$$= 1.43 \times 10^{-12}\ \text{m}^2\ \text{s}^{-1}$$

From the Einstein equation (Eq. 18.48) $\overline{x^2} = 2Dt$, and therefore are

$$t = \frac{(10^{-3}\ \text{m})^2}{2 \times 1.43 \times 10^{-12}\ \text{m}^2\ \text{s}^{-1}} = 3.5 \times 10^{5}\ \text{s}$$

15. The following values are to be inserted into Eq. 18.93

$R = 8.314$ J K^{-1} mol^{-1} $D = 5.96 \times 10^{-11}$ m^2 s^{-1}

$T = 393.15$ K $v_1 = 0.736$ cm^3 g^{-1}

$s = 4.6 \times 10^{-13}$ s $\rho = 0.998$ g cm^{-3}

Thus

$$M = \frac{8.314 \times 293.15 \times 4.60 \times 10^{-13}}{5.96 \times 10^{-11}\ (1 - 0.736 \times 0.998)}$$

$$= 70.86\ \text{J m}^{-2}\ \text{s}^2\ \text{mol}^{-1}$$

$$= 70.86\ \text{kg mol}^{-1} = 70\ 860\ \text{g mol}^{-1}$$

16. From Eq. 18.93

$$M = \frac{8.314 \text{ J K}^{-1} \text{ mol}^{-1} \times 298.15 \text{ K} \times 1.13 \times 10^{-12} \text{ s}^{-1}}{4.2 \times 10^{-11} \text{ m}^2 \text{ s}^{-1}(1 - 0.997/1.32)}$$

$$= 272.5 \text{ kg mol}^{-1} = 272\ 500 \text{ g mol}^{-1}$$

17. (a) From Eq. 18.86

$$f = \frac{(1 - V_1\rho)m\,\omega^2 x}{v} = \frac{(1 - V_1\rho)m}{s}$$

where $s \equiv v/\omega^2 x$ is the sedimentation coefficient (Eq. 18.87). The molecular mass m is

$$\frac{60\ 000 \text{ g mol}^{-1}}{6.022 \times 10^{23} \text{ mol}^{-1}} = 9.96 \times 10^{-23} \text{ kg}$$

The fractional coefficient is thus

$$f = \frac{(1 - 0.997/1.31) \times 9.96 \times 10^{-23} \text{ kg}}{4.1 \times 10^{-13} \text{ s}}$$

$$= 5.80 \times 10^{-11} \text{ kg s}^{-1}$$

(b) Volume of protein molecule is

$$\frac{9.96 \times 10^{-23} \text{ kg}}{1.31 \times 10^3 \text{ kg m}^{-3}} = 7.60 \times 10^{-26} \text{ m}^3$$

If the particle is spherical and its radius is r,

$$\frac{4}{3}\pi r^3 = 7.60 \times 10^{-26} \text{ m}^3$$

$$r = 2.63 \times 10^{-9} \text{ m}$$

According to Stokes's law (Eq. 18.77)

$$f = 6\pi\eta r$$

$$= 6\pi \times 8.937 \times 10^{-4} \text{ kg m}^{-1} \text{ s}^{-1} \times 2.63 \times 10^{-9} \text{ m}$$

$$= 4.43 \times 10^{-11} \text{ kg s}^{-1}$$

The fact that the observed f is greater than that calculated using Stokes's law may be attributed to the fact that the molecule is not spherical.

18. Limiting rate of sedimentation is given by Eq. 18.84:

$$v = \frac{(1 - V_1\rho)m\ g}{6\pi r\eta}$$

The mass of the particle is

$$m = \frac{4}{3}\pi(1.5 \times 10^{-7}\ \text{m})^3 \times 1.18 \times 10^3\ \text{kg m}^{-3}$$
$$= 1.668 \times 10^{-17}\ \text{kg}$$

Then

$$v = \frac{(1 - 0.998/1.18) \times 1.668 \times 10^{-17}\ \text{kg} \times 9.81\ \text{m s}^{-1}}{6\pi \times 1.002 \times 10^{-3} \times 1.5 \times 10^{-7}\ \text{kg s}^{-1}}$$
$$= 8.908 \times 10^{-9}\ \text{m s}^{-1}$$

The particle thus sediments a distance of $1\ \text{mm} = 10^{-3}\ \text{m}$ in

$$\frac{10^{-3}\ \text{m}}{8.908 \times 10^{-9}\ \text{m s}^{-1}} = 1.12 \times 10^5\ \text{s}$$

19. We can use Stokes's law to estimate the radius of the particles:

$$r = \frac{\mathbf{k}T}{6\pi\eta D} = \frac{1.381 \times 10^{-23} \times 298.15\ \text{J}}{6\pi \times 8.937 \times 10^{-4}\ \text{kg m}^{-1}\ \text{s}^{-1}}$$
$$= 1.2 \times 10^{-11}\ \text{m}^2\ \text{s}^{-1} = 2.036 \times 10^{-8}\ \text{m}$$

The mass of each particle is thus

$$m = \frac{4}{3}\pi(2.036 \times 10^{-8}\ \text{m})^3 \times 1.33 \times 10^3\ \text{kg m}^{-3}$$
$$= 4.70 \times 10^{-20}\ \text{kg}$$

The sedimentation coefficient is

$$s = \frac{(1 - V_1\rho)m}{f} = \frac{(1 - V_1\rho)m}{6\pi r\eta}$$

$$= \frac{(1 - 0.997 \times 1.33) \times 4.70 \times 10^{-20}\ \text{kg}}{6\pi \times 8.937 \times 10^{-4}\ \text{kg m}^{-1}\ \text{s}^{-1} \times 2.036 \times 10^{-8}\ \text{m}}$$

$$= 3.43 \times 10^{-11}\ \text{s}$$

20. (a) At 0°C

$$e^{-E/RT} = e^{-12\,600/8.314 \times 273.15} = 3.894 \times 10^{-3}$$

At 40.0°C

$$e^{-E/RT} = e^{-12\,600/8.314 \times 313.15} = 7.910 \times 10^{-3}$$

Viscosity at 40.0°C = $7.06 \times 10^{-4} \times \frac{3.894}{7.910}$ kg m^{-1} s^{-1}

$= 3.48 \times 10^{-4}$ kg m^{-1} s^{-1}

(b) At 0°C

$$(T/\text{K})^{-1.72}\, e^{543(T/\text{K})} = (273.15)^{-1.72}\, e^{543/273.15}$$

$$= 6.45 \times 10^{-5} \times 7.30 = 4.709 \times 10^{-4}$$

At 40.0°C

$$(T/\text{K})^{-1.72}\, e^{543(T/\text{K})} = (313.15)^{-1.72} \times e^{543/313.15}$$

$$= 5.097 \times 10^{-5} \times 5.66 = 2.89$$

Viscosity at 40.0°C = $7.06 \times 10^{-4} \times \frac{2.89}{4.709}$ kg m^{-1} s^{-1}

$= 4.33 \times 10^{-4}$ kg m^{-1} s^{-1}

The fluidity can be expressed as

$$\phi = A^{-1}(T/\text{K})^{1.72}\ e^{-4514\ \text{J mol}^{-1}/RT}$$

The activation energy is, by by definition

$$E = RT^2\ \frac{d\ \ln\ \phi}{dT}$$

$$\frac{d\ \ln\ \phi}{dT} = \frac{1.72\ \text{K}}{T} + \frac{4514\ \text{J mol}^{-1}}{RT^2}$$

$$= \frac{(1.72 \times 8.314 \times 293.15) + 4514\ \text{J mol}^{-1}}{RT^2}$$

$$= \frac{(4192 + 4514)\ \text{J mol}^{-1}}{RT^2} = \frac{8706\ \text{J mol}^{-1}}{RT^2}$$

The activation energy is thus

$$E = 8706\ \text{J mol}^{-1} = 8.7\ \text{kJ mol}^{-1}$$

21. By definition $E \equiv RT^2\ d\ \ln\ \phi^u/dT$

$$= -RT^2\ d\ \ln\ \eta^n/dT^u$$

From the emperical relationship, with

$$T^u = t^u + 273.15,$$

$$\ln\ \eta_t^u = \ln\ \eta_{20°}^u - \frac{a(T^u - 293.15) + b(T^u - 293.15)^2}{T^u - 164.15}$$

where $a = 1.370\ 23 \times 2.303 = 3.1556$

and $b = 8.36 \times 10^{-4} \times 2.303 = 1.925 \times 10^{-3}$

$$-\frac{d\ \ln\ \eta^u}{dT^u} = \frac{(T^u - 154.15)(a + 2bT^u - 586.3\ b) - a(T^u - 293.15) - b(T^u - 293.15)^2}{(T^u - 164.15)^2}$$

(a) At 20°C = 293.15 K the value of this is

$$\frac{129.0 \times (3.1556 + 1.129 - 1.128)}{16\ 641} = 0.0245$$

$$E = 0.0245 \times 8.314 \times 293.15^2 \text{ J mol}^{-1}$$

$$= 17.5 \text{ kJ mol}^{-1}$$

(b) At 100°C = 373.15 K the value is

$$\frac{(209.0 \times 3.46) - 252.5 - 12.32}{43\ 681} = 0.010\ 49$$

$$E = 0.010\ 49 \times 8.314 \times 373.15^2 \text{ J mol}^{-1}$$

$$= 12.1 \text{ kJ mol}^{-1}$$

The activation energy decreases as the temperature rises because of the breaking of hydrogen bonds between water molecules; the liquid becomes less structured with increase in temperature.

22. (a) In a hypothetical gas in which the molecules have no size there are no collisions and therefore no exchanges of momentum between the molecules. If there are no forces between the molecules, two layers can move past each other freely, and the viscosity is zero.

(b) If the molecules have no size but attract one another, a force is required to move one layer past another. The gas will therefore have a viscosity. Increasing the temperature will increase the molecular speeds and will decrease the viscosity, as in a liquid.

(c) If the molecules have no size but repel one another, a force will again be required to move one layer past another. There will again be a viscosity, which decreases with increasing temperature.

CHAPTER 19: WORKED SOLUTIONS
MACROMOLECULES

1. From Eq. 19.12 the polymerization rate is
$-\frac{d[\mathrm{M}]}{dt} = k_p\left(\frac{k_i}{k_t}\right)^{1/2}[\mathrm{M}]^{3/2}[\mathrm{C}]^{1/2}$ and the rate of initiation is $k_i[\mathrm{M}][\mathrm{C}]$ (Eq. 19.11). The chain length is therefore

$$\frac{k_p[\mathrm{M}]^{1/2}}{(k_ik_t)^{1/2}[\mathrm{C}]^{1/2}}$$

2. The rate of formation of CH_3 is $2I$. The steady-state equations are

$$2I - k_p[\mathrm{CH_3}][\mathrm{M}] - k_t[\mathrm{CH_3}]\Sigma[\mathrm{R}_n] = 0$$

$$k_p[\mathrm{CH_3}][\mathrm{M}] - k_p[\mathrm{CH_3CH_2CH_2}-][\mathrm{M}] - k_t[\mathrm{CH_3CH_2CH_2}-]\Sigma[\mathrm{R}_n] = 0$$

etc.

The sum of all the equations is

$$2I - k_t(\Sigma[\mathrm{R}_n])^2 = 0$$

so that

$$\Sigma[\mathrm{R}_n] = \left(\frac{2I}{k_t}\right)^{1/2}$$

The rate of removal of monomer is

$$v = k_p[\mathrm{M}]\Sigma[\mathrm{R}_n]$$

$$= k_p\left(\frac{2I}{k_t}\right)^{1/2}[\mathrm{M}]$$

3. $M_n = \frac{(10)(10\ 000) + (80)(20\ 000) + 10(40\ 000)}{100}$ g mol^{-1}

$= 21\ 000$ g mol^{-1}

$M_m = \frac{10(10\ 000)^2 + 80(20\ 000)^2 + 10(40\ 000)^2}{(10)(10\ 000) + (80)(20\ 000) + (10)(40\ 000)}$ g mol^{-1}

$= 23\ 333$ g mol^{-1}

4. $M_n = \dfrac{(5)(30\ 000) + (10)(60\ 000)}{5 + 10}$ g mol^{-1}

$= 50\ 000$ g mol^{-1}

$M_m = \dfrac{5(30\ 000)^2 + 10(60\ 000)^2}{5(30\ 000) + 10(60\ 000)}$ g mol^{-1}

$= 54\ 000$ g mol^{-1}

5. (a) If there are x mol with $M_r = 20\ 000$ there are x mol with $M_r = 30\ 000$:

$M_n = \dfrac{20\ 000\ x + 30\ 000\ x}{2x}$ g mol^{-1}

$= 25\ 000$ g mol^{-1}

$M_m = \dfrac{(20\ 000)^2\ x + (30\ 000)^2\ x}{(20\ 000)x + (30\ 000)x}$ g mol^{-1}

$= 26\ 000$ g mol^{-1}

(b) If there are x mol with $M_r = 30\ 000$ there are $1.5\ x$ mol with $M_r = 20\ 000$.

$M_n = \dfrac{1.5(20\ 000)x + (30\ 000)x}{2.5x}$ g mol^{-1}

$= 24\ 000$ g mol^{-1}

$= \dfrac{1.5(20\ 000)^2\ x + (30\ 000)^2\ x}{1.5(20\ 000)x + (30\ 000)x}$ g mol^{-1}

$= 25\ 000$ g mol^{-1}

6.

C/g dm^{-3}	3.30	6.40	9.25	12.50	14.90
π/Pa	27.4	53.8	78.5	107.5	129.5
πC^{-1}/Pa dm^3 g^{-1}	8.30	8.41	8.49	8.60	8.69

From a plot of πC^{-1} against c the extrapolated value of πC^{-1} is 8.175 Pa dm^3 g^{-1}

$= 8.175 \times 10^{-3}$ Pa m^3 g^{-1}

Therefore 8.175×10^{-3} Pa m^3 g^{-1} $= \frac{RT}{M}$

$= \frac{8.314 \times 300 \text{ J mol}^{-1}}{M}$

$M = 305\ 000$ g mol^{-1}

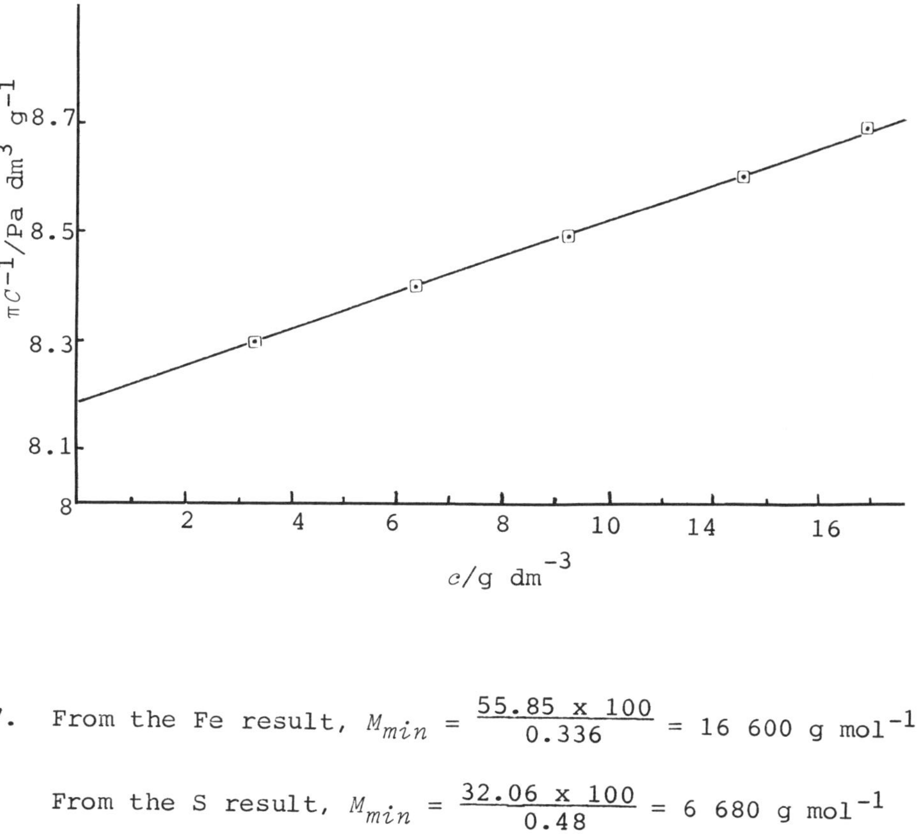

7. From the Fe result, $M_{min} = \frac{55.85 \times 100}{0.336} = 16\ 600$ g mol^{-1}

From the S result, $M_{min} = \frac{32.06 \times 100}{0.48} = 6\ 680$ g mol^{-1}

From the arginine result, $M_{min} = \frac{174.20 \times 100}{4.24}$

$= 4\ 100$ g mol^{-1}

The minimal molar mass consistent with these three values is $\approx 33\ 300$ g mol^{-1} (2 Fe, 5 S, 8 arginine).

8.

C/g dm^{-3}	4.52	9.37	16.0	23.7	29.8
π/Pa	246	572	1124	1940	2700
πC^{-1}/Pa dm^{-3} g^{-1}	54.4	61.0	70.3	81.9	90.6

From a plot of πC^{-1} against c the extrapolated value of πC^{-1} is 48.0 Pa dm^3 g^{-1}.

Therefore

$$48.0 \times 10^{-3} \text{ Pa m}^3 \text{ g}^{-1} = \frac{8.314 \times 293.15 \text{ J mol}^{-1}}{M}$$

$$M = 50\ 800 \text{ g mol}^{-1}$$

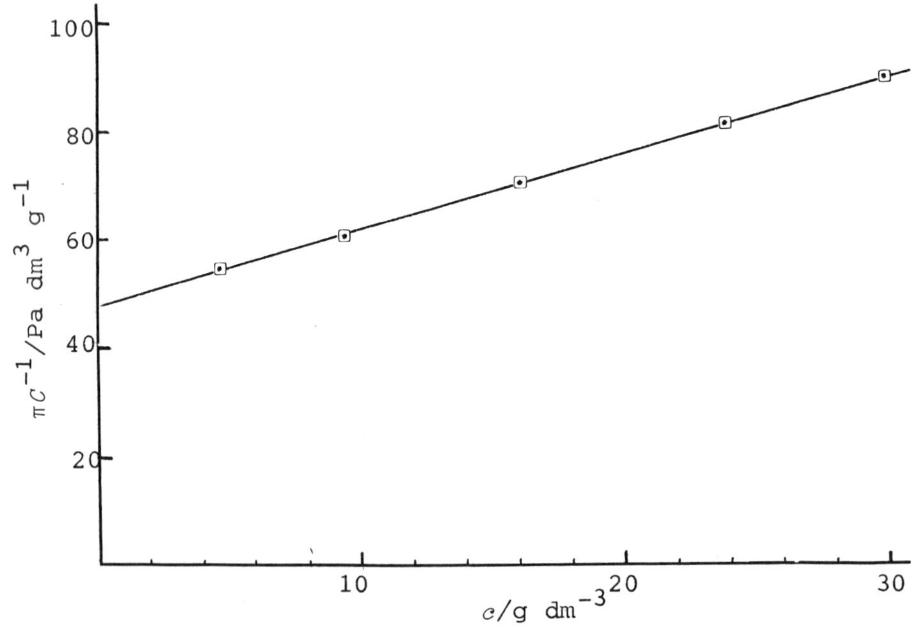

9. The relative molecular mass of ethylene is 28.05.

Therefore

$$N = \frac{50\ 000}{28.05} = 1783$$

The effective bond length is $\sqrt{2} \times 0.154 = 0.218$ nm

(a) $\overline{l^2} = 1783 \times 0.218^2 = 84.7\ \text{nm}^2$

$\sqrt{\overline{l^2}} = 9.21\ \text{nm}$

(b) $\bar{l} = \left(\dfrac{8 \times 1783}{3\pi}\right)^{1/2} 0.218 = 8.48\ \text{nm}$

(c) The distance between alternate carbon atoms

$= 2\sqrt{2} \times 0.154/\sqrt{3} = 0.251\ \text{nm}$

Fully-extended chain length

$= 0.5 \times 1783 \times 0.251 = 223.8\ \text{nm}$

10. The relative molecular mass of one styrene unit is 104.1. Therefore

$$N = \frac{75\ 000}{104.1} = 720$$

Effective bond length is $\sqrt{2} \times 0.154 = 0.218$ nm.

(a) $\overline{l^2} = 720 \times 0.218^2 = 34.2\ \text{nm}^2$

$\sqrt{\overline{l^2}} = 5.85\ \text{nm}$

(b) From Eq. 19.31

$$\lambda = \left(\frac{2 \times 720}{3}\right)^{1/2} 0.218 = 4.78\ \text{nm}$$

From Eq. 19.30, at 5 nm,

$$P = \frac{4}{\pi^{1/2} 4.78^3} \times 5^2\ e^{-5^2/4.78^2}\ \text{nm}^{-1}$$

$$= 0.174\ \text{nm}^{-1}$$

$dl = 0.02\ \text{nm}$

$P(l)dl = 0.174 \times 0.02 = 3.5 \times 10^{-3}$

(c) Most probable value of l is $\lambda = 4.78$ nm.

11. Force = (0.1 kg) x (9.81 m s^{-2}) = 0.981 N

The force, and therefore the mass required, is proportional to T. The mass required at 50°C is thus

$$(100 \text{ g}) \times \frac{323.15}{298.15} = 108.4 \text{ g}$$

12. For viscosity the mass average molar mass is used. The mass-average molar mass of the mixture is

$$M_w = \frac{10\ 000 + 100\ 000}{2} = 55\ 000 \text{ g mol}^{-1}$$

By the Mark-Houwink equation (Eq. 18.33)

$$[\eta] = k\ M_r^{\alpha}$$

or

$$\log_{10}[\eta] = \log_{10}k + \alpha \log_{10}M$$

For solution A

$$-3.21 = \log_{10}k + 4\alpha \qquad (1)$$

For solution B

$$-3.02 = \log_{10}k = 5\alpha \qquad (2)$$

For the mixture

$$x = \log_{10}k + 4.74\alpha \qquad (3)$$

From (1) and (2), $\alpha = 0.19$ and

$\log_{10}k = -3.21 - 4 \times 0.19 = -3.97$

$x = -3.97 + 4.74 \times 0.19 = -3.07$

$[\eta] = 10^{-3.07} = 8.5 \times 10^{-4}$ kg m^{-1} s^{-1}